AI Revolution in Chemistry

RSC Foundations

For a list of titles in this series see:
rsc.li/foundations

How to obtain future titles on publication:
A standing order plan is available for this series. A standing order will bring delivery of each new volume immediately on publication.

For further information please contact:
Book Sales Department, Royal Society of Chemistry, Thomas Graham House, Science Park, Milton Road, Cambridge, CB4 0WF, UK
Telephone: +44 (0)1223 420066, Fax: +44 (0)1223 420247
Email: booksales@rsc.org
Visit our website at books.rsc.org

AI Revolution in Chemistry

By

Brian McKew

Consultant, UK
Email: brian@mckew.co.uk

RSC Foundations No. 7

Paperback ISBN: 978-1-83707-215-6
PDF ISBN: 978-1-83707-216-3
EPUB ISBN: 978-1-83707-188-3
Print ISSN: 2978-1477
Electronic ISSN: 2977-0084

A catalogue record for this book is available from the British Library

The Royal Society of Chemistry is a charity, registered in England and Wales, Number 207890, and a company incorporated in England by Royal Charter (Registered No. RC000524), registered office: Burlington House, Piccadilly, London W1J 0BA, UK, Telephone: +44 (0)20 7437 8656.

For further information see our website at www.rsc.org

For general enquiries, please contact books@rsc.org

For EU product safety enquiries, please email books@rsc.org or contact Royal Society of Chemistry Worldwide (Germany) GmbH, Römischer Hof, Unter den Linden 10, 10117 Berlin.

Preface

Chemistry is a decision-making discipline. Every day, chemists choose which hypothesis to test, which analogue to make, which condition to try, which signal to trust, and which risk to avoid. In recent years—decisively since 2024—artificial intelligence (AI) has become a practical instrument for those decisions. It accelerates search, focuses experimentation, and turns scattered information into usable guidance. It is not a replacement for chemical judgement; it amplifies it when used with appropriate controls.

This book is written for practising chemists and adjacent scientists—medicinal, pharmaceutical, analytical, process, materials and formulation—who want a clear, execution-oriented view of AI. That includes colleagues working on active pharmaceutical ingredients (APIs), excipients (the ingredients other than the active drug), dosage forms, and the pharmaceutical formulations and processes that underpin them. The emphasis is chemistry-first: what to do, how to do it, how to validate it, and how to keep it safe, auditable and useful. Equations appear only when they change practice; governance appears whenever it protects people, product quality, or the environment.

You will find patterns that repeat across domains. Prediction, generation, planning and optimisation reappear whether you are ranking candidates, proposing routes, tuning a crystalliser, or screening materials. Representations matter: descriptors and fingerprints are strong baselines; graphs and sequences capture structure; 3D helps when shape or pose dominate; spectra and images are powerful when carefully pre-processed. Validation that predicts

RSC Foundations No. 7
AI Revolution in Chemistry
By Brian McKew
© Brian McKew 2026
Published by the Royal Society of Chemistry, www.rsc.org

deployment is non-negotiable: scaffold or time/site splits, calibrated uncertainty, and explicit applicability domains prevent disappointment in the lab or plant. And because chemistry is safety-critical, we make accessibility, permissions, data integrity and change control first-class concerns rather than after-thoughts.

The book is deliberately digital-first. Chapters are individually citable; figures include captions, alt-text and permissions tags; navigation uses labels rather than page numbers. British English and SI units are used throughout, with Royal Society of Chemistry style for references. Where the field is moving quickly—protein and materials foundation models; self-driving laboratories; regulatory guidance— we temper enthusiasm with evidence and point to durable workflows: retrieval-augmented assistants that cite sources; constraint-aware generators; Bayesian optimisation with feasibility and safety bounds; model cards and data cards that travel with systems through their lifecycle.

A note on scope. We do not attempt a comprehensive survey. Instead we curate workflows and exemplars that teams can copy and adapt in days to weeks, not months to years. Where claims rely on very recent literature, we cite cautiously and recommend orthogonal confirmation. Where images or schematics are third-party, we either request rights or provide original, non-infringing conceptual redraws.

Finally, a note of thanks. This work reflects the craft of many colleagues across chemistry, data science, automation, quality, and safety. Any errors are mine. I hope these pages help you make better, safer, faster decisions—and, as importantly, record why those decisions were sound.

Brian McKew

Acknowledgements

I would like to thank Professor Geoffrey D. Tovey for his generous mentorship and thoughtful guidance over many years. His rigorous thinking, deep experience in pharmaceutical sciences and constructive challenge have greatly influenced my approach to AI in chemistry and the way this book has been shaped.

RSC Foundations No. 7
AI Revolution in Chemistry
By Brian McKew
© Brian McKew 2026
Published by the Royal Society of Chemistry, www.rsc.org

About the Author

Brian McKew is an independent consultant in pharmaceutical sciences who works at the intersection of chemistry, data and artificial intelligence. He helps research and development teams explore and implement AI-enabled approaches to drug discovery, pharmaceutical formulation and dosage-form development, materials development and chemical manufacturing.

In his consulting practice, Brian focuses on turning abstract AI concepts into practical workflows, roadmaps and training that practising scientists can apply in real laboratory and plant environments. His goal is to bridge the gap between traditional experimental science and modern AI capabilities, enabling chemists and their organisations to make better, faster and more defensible decisions.

RSC Foundations No. 7
AI Revolution in Chemistry
By Brian McKew
© Brian McKew 2026
Published by the Royal Society of Chemistry, www.rsc.org

Glossary

Adverse outcome pathway (AOP)—A structured representation linking a molecular initiating event through key events to an adverse outcome at higher levels of biological organisation; used to contextualise *in vitro* and *in silico* evidence (Chapter 6).

Active pharmaceutical ingredient (API)—The chemically active substance in a medicine responsible for its therapeutic effect; in this book, AI workflows often focus on the design, optimisation and manufacture of APIs.

Excipients—The ingredients in a pharmaceutical formulation other than the active pharmaceutical ingredient (API), such as binders, disintegrants, fillers, coatings and stabilisers that help create the final dosage form.

Dosage form—The physical form in which a medicine is delivered, such as tablets, capsules, suspensions, solutions, injections or inhalers; also called the drug product.

Applicability domain (AD)—The region of chemical/process space in which a model's predictions are supported by training data; used to decide when to accept, qualify, or reject a prediction (Chapters 1, 6 and 7).

Bayesian optimisation (BO)—A sequential design strategy that uses a probabilistic surrogate model and acquisition function (*e.g.*, Expected Improvement, UCB) to select informative experiments under constraints (Chapters 2, 3 and 5).

Conformal prediction (CP)—A distribution-free framework that wraps any base model to produce valid prediction sets or intervals

RSC Foundations No. 7
AI Revolution in Chemistry
By Brian McKew
© Brian McKew 2026
Published by the Royal Society of Chemistry, www.rsc.org

with user-chosen error rates; enables per-sample uncertainty and reject options (Chapters 6 and 7).

Critical quality attribute (CQA)—A physical, chemical, biological, or microbiological property that must be within an appropriate limit to ensure product quality (Chapter 5).

Design of experiments (DoE)—Systematic planning of experiments (*e.g.*, factorial, fractional factorial, response surface) to model factor–response relationships efficiently (Chapters 2 and 5).

Digital twin—A computational representation of a process or system used for what-if analysis, soft sensing, controller testing, and scale-up planning; may be empirical or physics-informed (Chapter 5).

Graph neural network (GNN)—A deep learning architecture operating on graph-structured data (nodes/edges), widely used for molecular and materials property prediction (Chapters 1, 3 and 4).

High-throughput experimentation (HTE)—Parallelised screening of reaction or materials design spaces using miniaturised, standardised experiments (Chapter 2).

Model-predictive control (MPC)—An optimisation-based control strategy that computes future control moves subject to process models and constraints; often paired with soft sensors and RTO (Chapter 5).

Multi-parameter optimisation (MPO)—Simultaneous optimisation of multiple, often competing objectives (*e.g.*, potency, selectivity, solubility, safety) using composite desirability or Pareto methods (Chapters 3 and 4).

Physiologically based pharmacokinetic (PBPK) modelling—Mechanistic simulation of ADME processes across organs/tissues to predict concentration–time profiles; supports IVIVE and risk assessment (Chapter 6).

Process analytical technology (PAT)—Inline/online/at-line measurements (*e.g.*, IR, Raman, MS) and tools to design, analyse, and control manufacturing processes (Chapters 2 and 5).

Retrieval-augmented generation (RAG)—Integration of information retrieval with generative models so outputs are grounded in curated sources (Chapters 7 and 8).

Self-driving laboratory (SDL)—Closed-loop experimental platform combining automation, models (*e.g.*, BO, MPC), and sensors to propose –execute–measure–learn with safety envelopes (Chapters 2, 4 and 8).

Uncertainty quantification (UQ)—Estimating confidence in predictions or control actions (*e.g.*, *via* ensembles, Bayesian methods, CP); essential for governance and safe operation (Chapters 1, 2, 6 and 7).

Contents

RSC Foundations No. 7
AI Revolution in Chemistry
By Brian McKew
© Brian McKew 2026
Published by the Royal Society of Chemistry, www.rsc.org

2   Fundamentals of AI-driven Automation in Chemical Research   13

4　AI-enabled Materials Discovery and Optimisation　38

5 Transforming Chemical Manufacturing with AI   51

6 Enhancing Chemical Safety and Toxicology Through AI   59

7 Navigating Ethical Considerations and Regulations   72

10 Conclusion: Embracing the AI-driven Future of Chemistry   103

Further Reading   110

1 Understanding AI and Its Role in Chemistry

1.1 Chapter Overview

Artificial intelligence (AI) has moved from a curiosity on the periphery of the chemical sciences to an everyday instrument that informs how we frame hypotheses, design and prioritise candidates, plan syntheses, run experiments, and assure quality. This chapter lays the conceptual foundation for the remainder of the book. We unpack the principal families of AI you will encounter in practice—machine learning (ML), deep learning (DL), generative models, graph neural networks (GNNs), transformers, and Bayesian optimisation (BO)—but always through a chemistry-first lens. Rather than dwelling on equations, we emphasise what each method buys you at the bench, what it expects of your data, and how to validate its outputs so they are trustworthy.

Why now? In the last two years, several results have clarified AI's trajectory in chemistry. AlphaFold 3 extended structure prediction to biomolecular complexes (protein–ligand, nucleic acids, ions, post-translational modifications), enriching structure-guided design. AI-designed small molecules have progressed to peer-reviewed Phase 2a readouts, demonstrating that end-to-end pipelines—from target hypothesis to clinical signals—are achievable. Protein language models have generated functional proteins far from natural sequences, and self-driving laboratories have shown that model-guided, closed-loop experimentation can reach high-quality reaction conditions

RSC Foundations No. 7
AI Revolution in Chemistry
By Brian McKew
© Brian McKew 2026
Published by the Royal Society of Chemistry, www.rsc.org

while sampling only a small fraction of the search space. You need not become an AI specialist to benefit; you do need a clear sense of what these tools can and cannot do, how to provide fit-for-purpose data, and how to judge their answers.

By the end of this chapter, you will understand: (i) what "AI" means in practical chemical terms; (ii) how major model families map to common tasks; (iii) how to represent molecules, reactions, materials and protein complexes—including active pharmaceutical ingredients (APIs), excipients and dosage-form matrices—for learning; (iv) how to plan validation, uncertainty estimation, and applicability domain; and (v) how headline advances fit into robust chemistry workflows rather than standing as isolated demonstrations. Throughout, we emphasise three guardrails: synthesise-ability, process constraints, and prospective validation.

1.2 Core Concepts and Definitions (Chemistry-first)

1.2.1 What "AI" Means Here

At the bench level, AI is a family of statistical tools that learn patterns from data and support decisions—predicting a property (*e.g.*, solubility and yield), classifying a spectrum, generating plausible molecules or materials given a target profile, planning a synthesis route, or optimising conditions. The value proposition is twofold:

Scale and speed: once trained, a model evaluates candidates in milliseconds and can triage thousands to millions of options that would be impractical to simulate or assay.

Pattern discovery: models capture multivariable relationships that resist hand-crafted rules (*e.g.*, how subtle substructure changes shift permeability, or how composition and defects modulate a crystal's ionic conductivity).

AI is not a substitute for chemical judgement; it is an instrument—like NMR or a flow reactor—that amplifies capability when used with appropriate controls and can mislead if used without them.

1.3 Four Recurring Task Families

Predict—Learn a mapping from inputs to outputs.

Examples: pK_a, $\log P$, solubility, reaction yield/selectivity, band gap, chromatographic retention, spectral assignments.

Models: ridge/LASSO, random forests/gradient boosting, GNNs, transformers.

Generate—Propose new molecules/materials/protein sequences conditioned on desired properties or constraints.

Examples: small molecules with potency/selectivity/permeability and a specified building-block palette; crystal lattices targeting high Li^+ conductivity; enzyme variants compatible with a mechanistic hypothesis.

Models: variational autoencoders, diffusion models, sequence-to-sequence transformers; reinforcement learning can reward desirable designs.

Plan—Computer-aided synthesis planning (CASP): propose retrosynthetic disconnections, prioritise routes by cost/risk/greenness/precedent, and extract stepwise procedures usable by humans or robots.

Models: template-based and template-free retrosynthesis;[13] large-language-model (LLM)–assisted procedure extraction.

Optimise—Decide which experiments to run next to maximise information or performance with minimal experimentation.

Examples: reaction condition optimisation, catalyst loading–temperature trade-offs, formulation design.

Models: Bayesian optimisation with surrogate models; hybrid design-of-experiments.

These tasks frequently combine into a loop—Predict → Generate → Plan → Optimise—with laboratory data flowing back to update models (Figure 1.1).

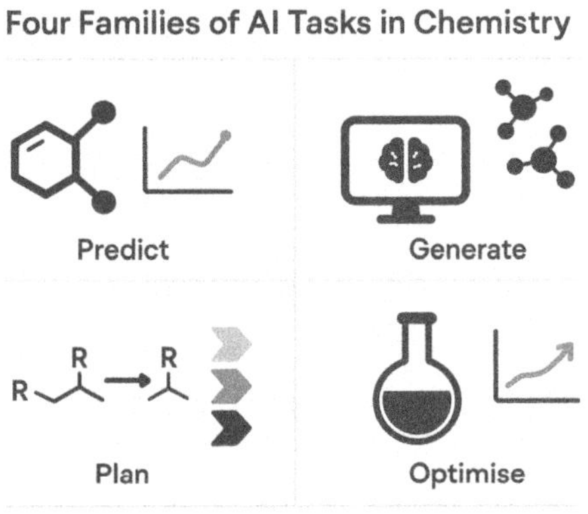

Figure 1.1 Four families of AI tasks in chemistry (Predict, Generate, Plan, Optimise).

1.4 Representations: Giving Chemistry to Models

A model can only learn from what you encode (Figure 1.2).

1.4.1 Molecules

- SMILES/SELFIES linearise structures; convenient for sequence models and high-throughput data handling.
- Graphs represent atoms (nodes) and bonds (edges); GNNs learn local environments and long-range correlations *via* message passing.[10]
- 3D conformers capture geometry and fields for shape-sensitive tasks (*e.g.*, binding and chiroptical properties).

1.4.2 Reactions

- Templates (atom-mapped rules) capture known disconnections.
- Template-free models map products → precursors directly.
- Action sequences (dose, ramp, stir, quench) support automation; they convert text protocols to machine-executable steps.

1.4.3 Materials

- Periodic crystal graphs or point clouds encode lattice geometry and composition.
- Compositional descriptors (*e.g.*, electronegativity and ionic radii) remain useful under data scarcity.

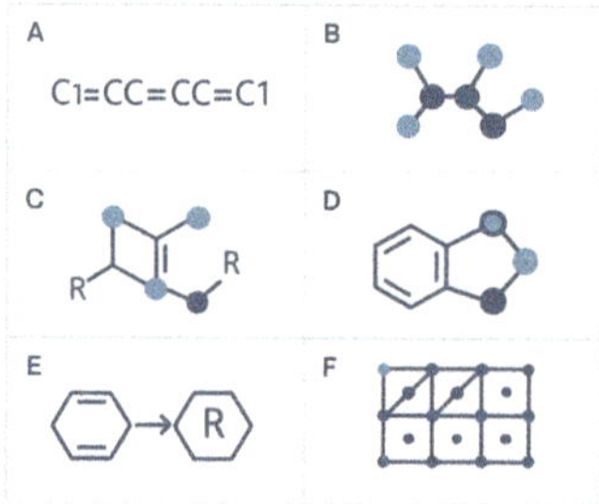

Figure 1.2 Representations used by models (SMILES/graphs/3D; reaction templates/actions; crystal graphs; protein/complex encodings).

1.4.4 Proteins and Complexes

- Sequences and 3D structures for individual chains.
- Complex graphs representing protein–ligand and protein–nucleic acid contacts enable modelling of assemblies.

Rule of thumb: start with the lightest representation that still contains the relevant signal; escalate to 3D or multimodal encodings only when the problem and data warrant it.

1.5 Model Families, Briefly

Baselines (ridge/LASSO; random forests; gradient boosting): Robust with small–medium datasets; interpretable; set an important performance floor.

Graph neural networks (GNNs): Strong defaults for molecular/ materials tasks;[9] learn chemistry without heavy feature engineering.

Transformers: Excellent for sequences and long-range dependencies (SMILES, proteins); backbone of LLMs.

Diffusion models: State-of-the-art generative models; iteratively denoise to structures satisfying constraints (molecules, crystals, even proteins).

Reinforcement learning (RL) and Bayesian optimisation (BO): RL nudges generation toward desired regions; BO chooses the next experiments, balancing exploration and exploitation.

1.6 Validation, Uncertainty, and Applicability Domain

Trust is earned, not assumed.

Splits that reflect deployment: For molecules, scaffold splits prevent over-optimistic metrics from look-alike analogues in train/test. For processes, temporal or site splits reflect forward generalisation and tech transfer.

Uncertainty quantification (UQ):[15] Ensembles and Monte-Carlo dropout offer pragmatic variance estimates; conformal prediction gives calibrated intervals.

Applicability domain (AD): Distance in learned embedding space (or to nearest neighbours) signals when an input is out-of-domain; avoid over-confident extrapolation (Figure 1.3).

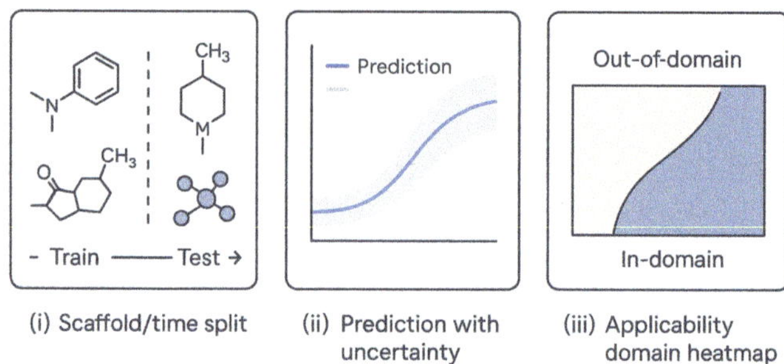

Figure 1.3 Validation and trust (splits, uncertainty, applicability domain).

Prospective validation: A small, well-chosen set of experiments (*e.g.*, 10–24 compounds or a modest optimisation batch) is worth more than a glossy retrospective AUC.

1.7 Methods and Workflows (Practical Detail)

1.7.1 The Minimum Information Set

Decision statement—one sentence describing the decision the model must support (*e.g.*, "Rank benchtop-suitable reagents and temperatures to maximise yield for a Buchwald–Hartwig coupling in flow.").

Outcome and metric—choose a metric that maps to risk or cost (*e.g.*, mean absolute error on yield; recall at a toxicity threshold; "within 2% of optimum").

Data inventory—sources (in-house assays, public databases, ELNs, instrument logs), sizes, and known biases; note units, SOPs, and missingness.[21]

Representation—select the simplest encoding that preserves signal[18] (descriptors/graphs/3D; reaction templates *vs* action sequences).

Baselines—always run simple baselines; they provide an honest lower bound and detect overfitting.

Validation plan—define scaffold/time/site splits; specify prospective tests up-front.

Governance—track seeds, code/library versions, hyperparameters, and data lineage; produce a one-page model card.

1.7.2 Property Prediction and Ranking

For physicochemical properties ($\log P$, pK_a), DMPK surrogates (permeability, clearance), and materials proxies (band gap), well-tuned baselines often suffice when data are clean and on-task. Where structure–property relationships are nonlinear or high-dimensional, GNNs or transformers deliver lifts[14]—provided the training data cover the relevant space. Avoid mixing assays with incompatible protocols or units; harmonise SOPs, normalise units, and document measurement uncertainty.

1.7.3 Generative Design Linked to Synthesise-ability

Generative models add the most value when tethered to practical chemistry.[8]

Conditioned objectives—combine potency/selectivity surrogates with ADME/tox alerts[16] and synthesise-ability[11] (*e.g.*, SCScore, building-block lists, reaction filters).

Route-aware triage—couple design outputs to CASP; prune multi-step or hazardous routes early; bake in green metrics (*e.g.*, *E*-factor targets, solvent preferences).

Materials constraints—constrain composition (scarce/toxic elements), space groups, and stability surrogates; favour families amenable to your synthesis assets.

Protein design—constrain to known folds/motifs; screen with structure prediction and biophysics before wet-lab investment.

1.7.4 Planning Syntheses and Writing Procedures

CASP systems propose disconnections and rank routes by step count, cost, risk, and precedent.[12] Procedure extraction (from publications or ELNs) translates prose to action lists[20] (dose, stir, heat, quench), supporting humans and robots. Always review and pilot: AI drafts are starting points, not gospel.

1.7.5 Optimising Reactions and Formulations

Bayesian optimisation (BO) is the default for efficient exploration.[6,19] Define factors (temperature, time, equivalents, catalyst/base, solvent, mixing/shear) and safety bounds; select a robust, automated readout (conversion, yield, selectivity, particle size). BO's surrogate model (Gaussian process or tree-based) and acquisition rule (*e.g.*, expected improvement) propose informative batches;[17] stop when gains plateau or an objective threshold is met. Use PAT (inline IR/Raman/MS) wherever possible; even a single reliable signal can enable closed-loop control.

1.7.6 Documentation and Reproducibility

Data sheets—schema, units, transformations, caveats; retain negative results.

Experiment logs—conditions; lot IDs; operator; instrument versions; anomalies.

Model cards—purpose; training data; metrics; UQ/AD; limitations; contact.

Change control—version model/data/code; record the rationale for any change to ranges, objectives, or algorithms.

1.8 Case Studies and Recent Signals (2024–2025)

1.8.1 Complex Structure Prediction as a Design Accelerant

AlphaFold 3 broadened structure prediction from single chains to biomolecular complexes,[1] yielding richer hypotheses for contact networks (*e.g.*, protein–ligand and protein–nucleic acid interactions). In practice, AF3 can prioritise series, suggest substituent directions, and motivate mutants to test binding hypotheses. It is not a substitute for experiment: pursue orthogonal evidence (co-crystal structures, NMR, mutational scans, SAR) to confirm or refute model-implied interactions. The payoff is fewer blind alleys and more discriminative experiments.

1.8.2 End-to-end Small-molecule Pipelines with Clinical Signals

Peer-reviewed Phase 2a results for an AI-designed small molecule[2] in idiopathic pulmonary fibrosis have demonstrated safety/ tolerability comparable to placebo and dose-dependent changes in lung function over 12 weeks. This does not imply universal success, but it does validate the premise that target discovery → generative design → CASP/synthesis → preclinical → clinical can be executed coherently. Operationally, teams should minimise cycle time, en- sure traceable hand-offs between modelling and experiment, and plan predefined decision gates.

1.8.3 Protein Language Models as Programmable Design Aides

Frontier protein LLMs that integrate sequence–structure–function[3] have generated working proteins far from natural sequence space (*e.g.*, new fluorescent proteins). For chemists, this points to near-term opportunities in chemo-enzymatic routes (designing or selecting enzyme variants to enable transformations) and in ligandable motif engineering. Again, treat outputs as hypotheses to be validated by structure and function assays.

1.8.4 Materials Discovery at Scale

Graph-based and diffusion methods (*e.g.*, large-scale crystal generators) have proposed hundreds of thousands of plausible materials;[4] autonomous labs have synthesised and characterised dozens within days to weeks. The implication is strategic: allocate more effort to down-selection and constraint-driven design up front, then run focused validation campaigns with fast iteration on composition and processing parameters.

1.8.5 Self-driving Labs and BO Templates[5]

Several 2024 studies showed closed-loop BO achieving >80% conversion[7] while sampling only a few per cent of a large condition space. Common ingredients: a calibrated surrogate with UQ; a well-bounded design space; an acquisition rule tuned to practical batch sizes; and crisp, automated readouts. These recipes are domain-agnostic, provided you can measure your objective reliably.

1.9 Common Pitfalls and How to Mitigate Them

1.9.1 "Great On Paper, Poor in the Lab"—Leakage and Non-causal Shortcuts

Fix: Use scaffold/time/site splits; remove trivial identifiers (*e.g.*, plate or operator codes) from features; perform error forensics to expose spurious correlates.

1.9.2 Out-of-domain Predictions

Fix: Quantify AD and UQ; refuse or flag low-confidence predictions; collect a few-shot stress set to harden the model where it will be used.

1.9.3 Ignoring Synthesise-ability and Process Constraints

Fix: Enforce building-block libraries, allowed reaction classes, step limits, and green constraints inside generation; triage with CASP; pilot early steps.

1.9.4 Over-trusting a Single Surrogate[22] (*e.g.,* Docking)

Fix: Use multi-objective scores (activity/selectivity $+$ ADME $+$ tox $+$ route $+$ risk) and require orthogonal evidence before committing to scale-up.

1.9.5 Inadequate Documentation and Governance

Fix: Maintain model cards, change logs, and dataset versioning; pre-register success criteria for prospective tests; define retrain triggers.

1.9.6 Accessibility and Permissions Gaps

Fix: Keep an alt-text register; tag figures as Original or Third-party (permission required); request rights well before submission.

1.10 Implementation Guidance (Checklists and Decision Frameworks)

1.10.1 First Deployment (Prediction/Ranking)

Define the decision and what action it changes.
   Audit data: provenance, units/SOPs, missingness/outliers, bias.
   Pick representation: descriptors/graphs/3D; justify escalation.
   Run baselines: document performance and failure modes.
   Train advanced model (GNN/transformer): add UQ/AD.
   Plan prospective test ($N = 10$–24): preregister metrics and thresholds.
   Operationalise: who consumes outputs, at what cadence, with what guardrails?
   Governance: model card; versioning; retrain policy.

1.10.2 Generative Campaign (Design → Make → Learn)

Brief: property windows; forbidden motifs; "must-have" substructures.

Constraints: building blocks; allowed reaction classes; step cap; green constraints; excluded reagents/solvents.

Scoring: multi-objective (activity/selectivity + ADME + tox + route + cost/risk).

Triage: route-aware down-selection; include a small "exploration" slice.

Pilot: synthesise a diverse, informative mini-set; capture failures.

Update: retrain predictors; refresh brief; iterate.

1.10.3 Closed-loop Optimisation (Reaction/ Formulation)

Design space: numeric/categorical factors; safety bounds; batch size; budget.

Metric: robust, automated readout (conversion/yield/selectivity; particle size; viscosity).

BO setup: surrogate with UQ; acquisition rule; stopping conditions.

Execution: integrate PAT; log every run (conditions, outcomes, anomalies).

Handover: package best conditions with robustness tests (*e.g.*, lot/ batch/operator variability).

1.11 Key Takeaways

AI is an instrument for chemical decision-making; its value depends on data quality, constraints, and validation.

Organise workflows around Predict–Generate–Plan–Optimise; close the loop so experiments improve models.

Start with baselines; escalate to GNNs/transformers/diffusion only when justified; report UQ and AD.

Tie design to synthesise-ability and process constraints; couple generation to CASP and green metrics.

Prefer prospective validation and closed-loop optimisation to purely retrospective benchmarks.

Maintain model cards, change logs, and dataset versioning; define retrain triggers and decision gates.

Recent advances (AF3, AI-designed clinical signals, protein LLMs, materials + robotics) boost capability but still demand orthogonal confirmation.

Prepare alt-text and permissions early—accessibility and rights are publication gatekeepers.

References

1. J. Jumper and D. Hassabis, *et al.*, Accurate structure prediction of biomolecular interactions with AlphaFold 3, *Nature*, 2024, **630**, 493–500.
2. Z. Xu, *et al.*, A generative AI-discovered TNIK inhibitor for idiopathic pulmonary fibrosis: randomised, placebo-controlled Phase 2a study, *Nat. Med.*, 2025, **31**, 2602–2610.
3. T. Hayes, *et al.*, ESM3: a frontier model of sequence–structure–function; *de novo* fluorescent protein design, *Science*, 2025, **387**, 850–858.
4. A. Merchant, *et al.*, Scaling deep learning for materials discovery, *Nature*, 2023, **624**, 80–85.
5. G. Tom, *et al.*, Self-Driving Laboratories for Chemistry and Materials Science, *Chem. Rev.*, 2024, **124**, 9633–9732.
6. Y. Wu, A. Walsh and A. M. Ganose, Race to the bottom: Bayesian optimisation for chemical problems, *Digital Discovery*, 2024, **3**, 1086–1100.
7. O. Schilter, *et al.*, Combining Bayesian optimisation and automation to optimise reaction conditions and routes, *Chem. Sci.*, 2024, **15**, 7732–7741.
8. D. C. Elton, Z. Boukouvalas, M. D. Fuge and P. W. Chung, Deep learning for molecular design—a review of the state of the art, *Mol. Syst. Des. Eng.*, 2019, **4**, 828–849.
9. D. Duvenaud, D. Maclaurin, J. Iparraguirre, R. Bombarell, T. Hirzel, A. Aspuru-Guzik and R. P. Adams, Convolutional networks on graphs for learning molecular fingerprints, Advances in Neural Information Processing Systems 28 (NIPS 2015)
10. J. Gilmer, S. S. Schoenholz, P. F. Riley, O. Vinyals and G. E. Dahl, Neural message passing for quantum chemistry, *Proc. Mach. Learn. Res.*, 2017, **70**, 1263–1272.
11. B. Sanchez-Lengeling and A. Aspuru-Guzik, Inverse molecular design using machine learning: generative models, *Science*, 2018, **361**, 360–365.
12. C. W. Coley, *et al.*, Computer-assisted synthesis planning: status and challenges, *Acc. Chem. Res.*, 2018, **51**, 1281–1289.
13. M. H. S. Segler, *et al.*, Planning chemical syntheses with deep neural networks and symbolic AI, *Nature*, 2018, **555**, 604–610.
14. K. Yang, *et al.*, Analyzing learned molecular representations for property prediction, *J. Chem. Inf. Model.*, 2019, **59**, 3370–3388.
15. J. P. Janet, *et al.*, A quantitative uncertainty metric controls error in neural-network predictions of chemical properties, *Chem. Sci.*, 2019, **10**, 7913–7922.
16. A. Mayr, G. Klambauer, T. Unterthiner and S. Hochreiter, DeepTox: toxicity prediction using deep learning, *Front. Environ. Sci.*, 2016, **3**, 80.
17. W. H. Green, *et al.*, Data and uncertainty in chemical kinetics and catalysis, *ACS Catal.*, 2021, **11**, 10672–10696.
18. G. Landrum, RDKit: Open-source cheminformatics, *J. Cheminf.*, 2013.
19. P. Frazier, A tutorial on Bayesian optimisation, *arXiv*, 2018, arXiv:1807.02811.
20. P. Schwaller, *et al.*, Predicting retrosynthetic disconnections and reaction conditions with transformers, *Mach. Learn.: Sci. Technol.*, 2021, **2**, 015016.
21. Z. Wu, B. Ramsundar, E. N. Feinberg, J. Gomes, C. Geniesse, A. S. Pappu, K. Leswing and V. Pande, MoleculeNet: a benchmark for molecular machine learning, *Chem. Sci.*, 2018, **9**, 513–530.
22. A. Bender and M. Cortés-Ciriano, Artificial intelligence in drug discovery: what is realistic, and what are the pitfalls?, *Drug Discovery Today*, 2021, **26**(2), 511–524.

2 Fundamentals of AI-driven Automation in Chemical Research

2.1 Chapter Overview

Automation has always been part of the chemist's craft, from the first reflux condensers to multichannel pipettes and programmable synthesisers. What is new is autonomy: the purposeful coupling of machine learning (ML) with robotics and process analytical technology (PAT) so that experiments can be planned, executed, measured, and iterated with minimal human intervention. This chapter builds the conceptual and practical scaffolding for that shift. We define the core elements of an AI-driven laboratory—high-throughput experimentation (HTE), closed-loop optimisation, digital twins, soft sensors, and self-driving laboratories (SDLs)—and position them within typical chemical workflows: reaction discovery and optimisation, formulation design, materials screening, and structure-guided experimentation.

The motivation is not fashion but precision, speed, reproducibility, and safety. HTE and closed-loop optimisation can compress month-long condition screens into days, identify non-obvious regimes (*e.g.*, low-loading/high-selectivity conditions), and improve repeatability by keeping instruments, environments, and protocols under tighter control. In drug discovery and chemical biology, upstream AI (such as AlphaFold 3 and protein language models) proposes testable structural hypotheses;[19] automation then executes the hundreds of

RSC Foundations No. 7
AI Revolution in Chemistry
By Brian McKew
© Brian McKew 2026
Published by the Royal Society of Chemistry, www.rsc.org

targeted variants and conditions needed to validate, refine, and exploit those hypotheses. In materials research, model-guided robotics traverse compositional and process spaces far larger than any manual study, while enforcing safety limits and logging every step for later analysis.

Our objective is to bridge the concepts and the kit. We clarify how AI and automation interact in practice; how to encode design variables and constraints; how to choose between design-of-experiments (DoE) and Bayesian optimisation (BO); how to specify and maintain a data and metadata spine; and how to stage your first autonomy project so that it is tractable, auditable, and genuinely useful. Real-world case studies from 2024–2025 show what has worked, why it worked, and where prudence is still required.

2.2   Core Concepts and Definitions

2.2.1   Automation *Versus* Autonomy

Automation executes a pre-specified sequence of actions—dilutions, transfers, temperature ramps, chromatographic runs—reducing human variability and freeing attention.

Autonomy adds decision-making: a model proposes which experiments to run next, given results so far, often with a formal acquisition strategy (*e.g.*, BO). The lab becomes a closed loop: Design → Make → Measure → Learn → Redesign (Figure 2.1).

2.2.2   High-throughput Experimentation (HTE)

HTE breaks large experimental questions into many small, parallel, standardised units (*e.g.*, 24–1536 wells; multi-channel

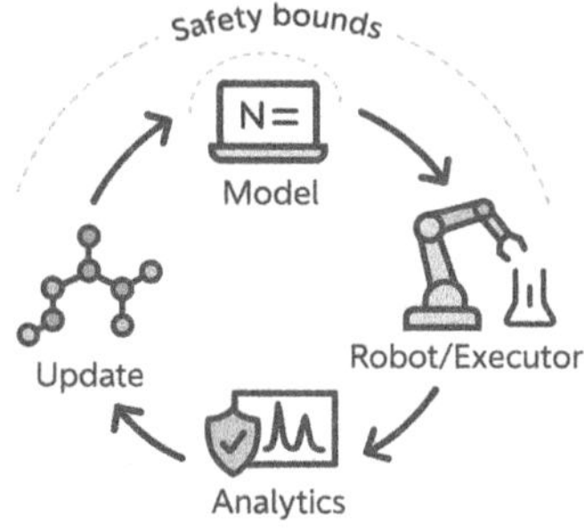

Figure 2.1   An AI-driven closed loop for reaction optimisation.

flow manifolds; combinatorial thin-film arrays). Two choices dominate HTE success:

The design space: catalysts/ligands/bases/solvents, temperature ($^\circ$C), time (min), equivalents ($\mathrm{mol\,mol^{-1}}$), pH, residence time (s), light intensity ($\mathrm{mW\,cm^{-2}}$), reactor configuration.

The sampling scheme: from coarse grids and fractional factorials for screening to space-filling designs (*e.g.*, Latin hypercube) that seed a surrogate model for subsequent BO.

HTE may be open-loop (preplanned matrices) or closed-loop (iteratively selected by a model using current results).

2.2.3 DoE *vs.* BO (and Hybrids)

DoE structures exploration is efficient when the number of factors is modest, and the response is approximately smooth/low-order. It is ideal for screening and interaction mapping, providing interpretable main effects and interaction plots that chemists can rationalise mechanistically.

BO substitutes a surrogate model[5,13] (*e.g.*, Gaussian process, random forest, gradient-boosted trees) for the unknown response surface and uses an acquisition function[6] (*e.g.*, expected improvement, upper-confidence bound, Thompson sampling) to choose new experiments that optimally trade exploitation and exploration. This is effective when experiments are expensive, when variables are both continuous and categorical, and when constraints matter (temperature ceilings, solvent compatibility).

Hybrid strategies are common: begin with a space-filling DoE to seed the surrogate, then hand over to BO for rapid convergence. For multi-objective problems (*e.g.*, maximise yield and selectivity while minimising *E*-factor), use Expected Hypervolume Improvement (EHVI) or Pareto-front navigators (Figure 2.2).

DoE vs Bayesian optimisation

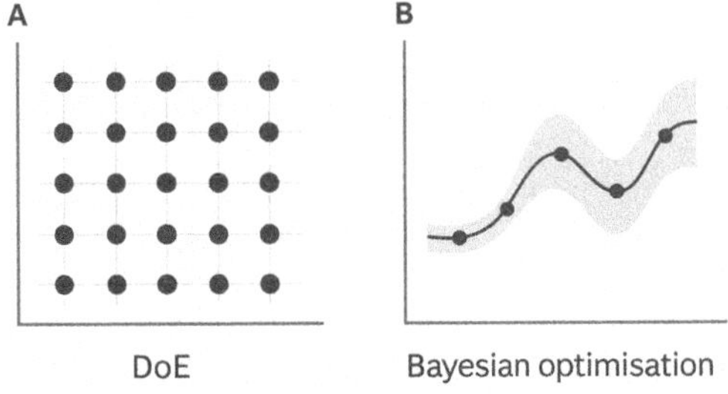

Figure 2.2 DoE *versus* Bayesian optimisation.

2.2.4 Process Analytical Technology (PAT) and Soft Sensors

PAT provides inline/online/at-line measurements that turn a batch or flow system into an observable dynamical system. Common PAT modalities include IR, Raman, UV–vis, MS, NMR (flow-through), calorimetry, particle-size analysers, and conductivity/density probes. PAT places numerics where intuition used to sit, enabling real-time feedback and richer datasets.

Soft sensors are ML models[15] that infer hard-to-measure or delayed variables—conversion, viscosity, number-average molecular mass $\bar{M}n$, particle-size distribution—from PAT time-series and actuator histories. They provide the "missing eyes and ears" required for closed-loop control when direct sensors are unavailable or impractical.

2.2.5 Digital Twins

A digital twin is a living computational representation of your process or assay: anything from an empirical surrogate (*e.g.*, a Gaussian-process map of yield *vs.* factors) to a hybrid model that combines first-principles balances with data-driven corrections (*e.g.*, unmodelled fouling or non-ideal mixing). Twins enable what-if analysis, controller design (including model-predictive control, MPC), and can serve as low-risk training grounds for reinforcement-learning policies.

2.2.6 Self-driving Laboratories (SDLs)

An SDL is not a single machine but an orchestrated capability:[7] models (predictors, surrogates) + acquisition logic (BO/MPC) + scheduling + robotics (liquid handling, flow, solid handling) + PAT + data/metadata capture + safety interlocks. Teams often start with a minimal loop—say, a liquid handler, a plate reader, and a BO script—then add flow skids, inline analytics, multi-objective acquisition, and remote operation. The common thread is the closed loop: propose $\rightarrow$ execute $\rightarrow$ measure $\rightarrow$ learn.[18]

2.3 Methods and Workflows

2.3.1 Designing an Autonomy-ready Experiment

Clarify the decision. Are you aiming to discover any active condition, to maximise a target metric (*e.g.*, $\geq$90% isolated yield), or to map a response surface for mechanistic understanding? The answer sets the resolution, the number of variables, and the choice of search strategy (DoE *vs.* BO).

Define variables and bounds. Enumerate numeric and categorical factors; set hard safety limits up front (temperature, pressure, exotherm thresholds, solvent incompatibilities). Consider throughput and analysis bottlenecks to pick a realistic number of factors and levels.

Choose a sampling scheme.[12]

DoE for broad screening (*e.g.*, fractional factorial in 96-well plates).

BO for targeted optimisation once seed data exist (*e.g.*, 20–60 experiments).

Hybrid: start with a space-filling seed (Latin hypercube) to initialise the surrogate.

Plan replication and randomisation: include replicates (5–20% of runs) to estimate noise/drift; randomise run order and plate layout to mitigate confounding (edge effects, diurnal drift, warm-up). Define controls and calibration standards (*e.g.*, internal standard for HPLC response factor normalisation).

Instrument considerations: ensure pipetting accuracy over the viscosity range; quantify and minimise dead volume and carry-over; verify irradiance at the reaction plane for photoredox; characterise residence-time distributions and mixing in flow microreactors; confirm PAT linear range and response times.

Metadata and identifiers: Assign a globally unique identifier to each experiment; log reagent lot/age, solvent batch, operator, instrument firmware, ambient conditions, and timestamps for all actions (aspirate/dispense/heat/measure). Adopting FAIR/ALCOA+ practices early accelerates modelling and eases audit.

2.3.2 Representing Experiments for Learning

Build a tidy table with one row per experiment:

Design variables: numerical (scaled/normalised), categorical (one-hot or learned embeddings).

Context variables: batch, plate/well indices, reactor channel, instrument ID (use for diagnostics; exclude from training unless justified to avoid leakage).

Outcomes: primary objectives (yield, selectivity, particle size), secondary metrics (impurities, *E*-factor), and uncertainty (replicate variance, instrument error).

Constraints/flags: precipitation, clogging events, safety aborts, QC failures.

For BO, maintain clean X (factors) and y (objective) matrices; for multi-objective BO, store vector outcomes with consistent scaling and direction (maximise *vs.* minimise). For classification (*e.g.*, active/ inactive), record decision thresholds and class balance.

2.3.3 Choosing and Configuring Bayesian Optimisation

Surrogate models.[9]

Gaussian processes (GPs): strong default for low-to-moderate dimension, smooth responses; built-in uncertainty.

Random forests/gradient-boosted trees: robust to rugged, non-smooth landscapes; consider quantile regression or NGBoost for uncertainty.

Bayesian neural networks/deep kernel learning: for higher-dimensional or non-stationary problems when data are sufficient.

Acquisition functions.

Expected Improvement (EI): robust single-objective default.

Upper Confidence Bound (UCB): explicit exploration parameter.

Thompson Sampling: simple, effective for batches.

EHVI/Pareto: for multi-objective trade-offs (yield–selectivity– *E*-factor–cost).

Batching and constraints.

For batch execution, use batch EI or q-EI and enforce diversity (*e.g.*, maximising minimum inter-point distance in the factor space).

Encode hard constraints[10] (temperature, pressure, pH, solvent compatibility) inside the acquisition. Add feasibility models (binary classifiers) trained on historical outcomes to avoid unproductive or unsafe regions.

Stopping rules.

Convergence (no improvement >1–2% over k batches)

Budget (N experiments or time limit)

Satisficing (stop at a target threshold; pivot to robustness tests)

Noise, replication, robustness.

Use replicate-aware surrogates or add observation noise terms; regularly re-measure top candidates.

Around the recommended optimum, perturb conditions ($\pm\Delta T$, $\pm10\%$ equivalents, alternative solvent grade, different operator) to estimate robustness bands.

2.3.4 Building a Minimal Self-driving Loop

A minimal viable loop comprises the following:

Planner: a Python microservice hosting the surrogate and acquisition logic; communicates *via* CSV/JSON/REST or a message bus.

Executor: a liquid handler or flow controller with a queue API; recipe templates (plate maps, reagent definitions, set-points) defined in version-controlled files.

Sensor: an inline or at-line readout (UV–vis/IR plate reader, HPLC autosampler, inline IR/Raman/MS).

Orchestrator: a scheduler or lightweight state machine managing propose → execute → measure → learn cycles, with retries and safe-halt states.

Data layer: a small database or ELN-backed store for X, y, metadata, and raw artefacts (chromatograms, spectra), all versioned.

Safety: interlocks (temperature/pressure/UV), watchdogs for leaks/clogs/over-current, physical *E*-stops, and permissions that limit model authority during early phases.

Start with one objective on one instrument. Demonstrate stability and value, then expand to multi-instrument orchestration, multi-objective BO, and remote operation.

2.3.5　Soft Sensors and Hybrid Control[14]

Where PAT is indirect, train soft sensors to infer target variables (conversion, viscosity, $\bar{M}n$, particle size) from spectra/time-series. Establish calibration protocols (blanks; substrate/product standards; periodic drift checks) and re-calibrate after maintenance (*e.g.*, lamp replacement). For control, combine classical controllers (PID/MPC) with ML-assisted set-point suggestion; avoid end-to-end "black-box" control until you can demonstrate safety and robustness (Figure 2.3).

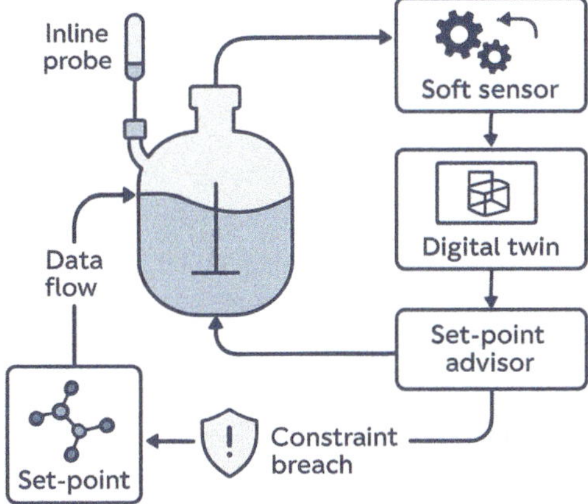

Figure 2.3　Digital twin and soft-sensor integration with PAT.

2.3.6 Digital Twins for Planning and Scale-up

A twin can be:

Empirical/surrogate: *e.g.*, a GP or gradient-boosted model approximating yield *vs.* factors.

Hybrid: first-principles balances + kinetic models + ML corrections for unmodelled effects (fouling, non-ideal mixing).

Multi-fidelity: low-fidelity simulators with occasional high-fidelity experiments, stitched *via* co-kriging or deep ensembles.

Use twins to plan solvent swaps, residence-time changes, parallelisation, and what-if analyses. Keep twins parsimonious; favour variables you can actually measure and control.

2.4 Case Studies and Recent Signals (2024–2025)

2.4.1 Self-driving Laboratories: From Prototype to Platform

A 2024 *Chemical Reviews* survey synthesised the state of SDLs[3] across synthesis, catalysis, and materials discovery, highlighting repeated patterns: modular hardware; a robust data/metadata spine; surrogate-based BO with batch-capable acquisition; and careful safety interlocks and operational discipline. Teams reported $10\times - 100\times$ throughput gains *versus* manual optimisation and improved reproducibility from standardised handling and inline QC. A key theme was that SDL success depends as much on people and process—calibration discipline, maintenance cadence, clear ownership, and comprehensive logging—as on algorithmic sophistication.

2.4.2 Closed-loop Optimisation with Minimal Experiments

A 2024 *Nature Chemistry* paper showed sequential, closed-loop BO[1] attaining near-optimal performance while sampling only ~2–3% of a ~4500-point design space in photocatalytic and metallophotocatalytic contexts. The loop paired a Gaussian-process surrogate with batch expected improvement, seeded by a space-filling design and constrained by explicit safety bounds. Critically, the team incorporated replicates and robustness checks (*e.g.*, perturbing conditions around the recommended optimum), which should be treated as standard practice.

2.4.3　Joint Condition and Route Discovery

A *Chemical Science* (2024) study combined BO with automated synthesis[2] to co-optimise reaction conditions and route variants, reporting >80% conversion after sampling only ~0.2% of candidates. Success depended on (i) a surrogate that handled both categorical (ligands/ solvents) and numeric factors; (ii) feasibility filters (*e.g.*, avoid hazardous combinations); and (iii) a well-engineered metadata pipeline so the optimiser trusted the measurements,[20] including failures.

2.4.4　Upstream Structure Models Guiding Downstream Autonomy

AlphaFold 3 (2024) introduced diffusion-based joint modelling[4] of protein–ligand, protein–DNA/RNA, and ion complexes with improved local accuracy. Chemical biology teams use AF3 to prioritise mutants, ligand series, and buffer conditions that probe specific interactions; autonomy reduces the cost of running the resulting matrices. Protein language models (*e.g.*, ESM-class) propose sequence variants with desired stability or binding propensities; automation closes the loop between design and assay.[17] The benefit is not perfect *a priori* prediction, but faster, better-targeted experimentation.

2.4.5　Materials Discovery at Scale

Large-scale graph and diffusion models proposed hundreds of thousands of plausible crystalline materials; "A-Lab"-style autonomous platforms[8] synthesised and evaluated dozens within days to weeks. Typical pipelines front-load composition and space-group filters, then use BO over processing parameters (temperature ramp, anneal time, precursor ratio) to converge rapidly on target properties (*e.g.*, ionic conductivity). Strategically, you prune with models, validate with robots, and tighten with BO.

2.5　Common Pitfalls and How to Mitigate Them

2.5.1　Confounding and Drift

Risk: Plate-edge effects, evaporative losses, reagent ageing, lamp intensity drift, or channel-to-channel variation masquerade as chemistry.

Mitigation: Randomise placements/run order; include replicates and standards; monitor PAT baselines; schedule calibration/maintenance; track reagent lots/age.

2.5.2 Data Leakage and Optimistic Metrics

Risk: Training on plate ID, operator or timestamp yields artefactual performance; naive CV ignores correlation structure.

Mitigation: Use scaffold/time/site splits; reserve operational IDs for diagnostics; require external/prospective validation.

2.5.3 Acquisition Myopia

Risk: BO over-exploits a local basin; batch points cluster; global optima missed.

Mitigation: Increase exploration (UCB parameter); enforce batch diversity; interleave space-filling batches; use multi-fidelity BO.

2.5.4 Objective Mis-specification

Risk: Optimising surrogate readouts that do not translate to isolated yield, stability or scale-up feasibility.

Mitigation: Align objectives with scientific/business goals; add penalties (impurities, E-factor, cost); verify top hits with orthogonal assays and scale-out tests.

2.5.5 Over-automation Without Safety

Risk: Models command heaters/pumps/lasers without robust interlocks; HAZOP not performed.

Mitigation: Implement hardware/software interlocks, watchdogs, E-stops; limit model authority early; conduct HAZOP; define "safe-halt" states.

2.5.6 Under-specified Metadata and Governance

Risk: Missing units/SOPs/timestamps; no failure logging; untraceable lineage.

Mitigation: Enforce ELN/LIMS schemas; adopt ALCOA+; version datasets/models; store raw data (chromatograms/spectra), not just aggregates.

2.5.7 Over-promising from Upstream Models

Risk: Treating AF3 or LLM suggestions as ground truth; ignoring uncertainty/alternatives.

Mitigation: Use AF3/LLMs for prioritisation, not final inference; design discriminative experiments; seek structural/biophysical confirmation.

2.6 Implementation Guidance (Checklists and Decision Frameworks)

2.6.1 Launch Your First HTE Screen[11]

Frame the question (*e.g.*, "Which of 60 catalysts/ligands yields $\geq 70\%$?").

Select variables: 3–6 factors $\times$ 2–4 levels; set safety bounds.

Choose design: Fractional factorial or Latin hypercube (48–96 runs) to seed models.

Controls and standards: Blanks; positives; internal standards for quantification.

Automate layout: Plate/flow maps with randomised assignments; verify pipetting and mixing.

Run with PAT: Capture inline/at-line readouts; predefine QC thresholds and salvage rules.

Analyse: Main-effects/interaction plots; shortlist conditions/factors for BO.

2.6.2 Transition from HTE to Closed Loop (BO)

Seed the surrogate with cleaned HTE data.

Pick acquisition (EI/UCB/Thompson); set batch size (8–16 typical).

Encode constraints ($T \leq 120$ °C; no peroxides with strong base; [A]/[B] bounds; maximum back-pressure).

Plan replication: 10–20% repeats for drift detection (current best + random sentinel).

Run–measure–learn: Update surrogate; recompute acquisition; repeat until convergence/budget/satisficing.

Confirm: replicate best conditions at scale; test robustness (lot/operator/day); store robustness bands.

2.6.3 Minimal Self-driving Loop Kit

Hardware: $1\times$ liquid handler or flow skid; plate reader or HPLC/GC; optional inline IR/UV or MS; temperature control; stir/shake/irradiate modules.

Software: Python (BoTorch/GPy/Scikit-Optimise) for BO; instrument APIs or CSV/REST bridges; lightweight scheduler/state machine.

Data/metadata: Experiment DB; ELN integration; schema for factors/outcomes/metadata/artefacts; consistent IDs; versioning.

Safety: Interlocks; watchdogs; *E*-stop; spill/leak detection; eye-safe enclosures for photochemistry; pressure relief.

2.6.4   RL or MPC Instead of BO?

Reinforcement learning (RL)[16] for sequential control with delayed/partial rewards (*e.g.*, crystalliser or bioreactor control under non-stationary feed). Requires simulators/twins and stringent safety constraints.

Model-predictive control (MPC) for continuous processes with dynamics and constraints; can incorporate ML surrogates (hybrid MPC) while enforcing hard bounds.

2.6.5   Data Governance Checklist

Provenance: Who/when/where; code/model/SOP versions; instrument firmware.

Confidentiality and integrity: Access control; ALCOA+; audit trails.

Retention: Keep raw signals (chromatograms/spectra), not just aggregates.

Change control: Register any change to design space, objectives, hardware, or software.

Model card: Purpose, data, metrics, UQ/AD, limitations, retrain triggers.

2.7   Key Takeaways

Autonomy = automation + intelligence: Closed loops connect models, robots, and PAT to design, run, and refine experiments.

Start simple: A minimal loop (one objective, one instrument) delivers large gains if the data/metadata spine and safety interlocks are sound.

Choose the right search strategy: DoE for screening; BO for efficient, constrained optimisation; use multi-objective BO for yield–selectivity–*E*-factor trade-offs.

Front-load constraints: Encode safety, synthesise-ability, cost, and green metrics inside the optimiser.

Trust *via* validation: Replicate; quantify uncertainty and applicability domain; perform prospective tests and robustness checks.

Soft sensors and digital twins turn limited measurements into actionable state estimates and enable "what-if" planning.

People and process matter: Calibration, metadata hygiene, change control, and governance are as critical as algorithms.

Upstream AI (AF3, protein LLMs) and scaled materials platforms guide what to test; autonomy makes systematic testing feasible.

References

1. X. Li, *et al.*, Sequential closed-loop Bayesian optimisation for chemical discovery and optimisation, *Nat. Chem.*, 2024, **16**, 1286–1294.
2. O. Schilter, *et al.*, Combining Bayesian optimisation and automation to simultaneously optimise reaction conditions and routes, *Chem. Sci.*, 2024, **15**, 7732–7741.
3. G. Tom, *et al.*, Self-Driving Laboratories for Chemistry and Materials Science, *Chem. Rev.*, 2024, **124**, 5905–5977.
4. J. Jumper and D. Hassabis, *et al.*, Accurate structure prediction of biomolecular interactions with AlphaFold 3, *Nature*, 2024, **627**, 795–803.
5. P. I. Frazier, A Tutorial on Bayesian Optimisation, *arXiv*, 2018, arXiv:1807.02811.
6. B. Shahriari, K. Swersky, Z. Wang, R. P. Adams and N. de Freitas, Taking the Human Out of the Loop: A Review of Bayesian Optimisation, *Proc. IEEE*, 2016, **104**, 148–175.
7. F. Häse, L. M. Roch and A. Aspuru-Guzik, Next-Generation Experimentation with Self-Driving Laboratories, *Trends Chem.*, 2019, **1**, 282–291.
8. B. P. MacLeod, *et al.*, Self-Driving Laboratory for Accelerated Discovery of Thin-Film Materials, *Sci. Adv.*, 2020, **6**, eaaz8867.
9. F. Häse, *et al.*, Phoenics: A Bayesian Optimizer for Chemistry, *ACS Cent. Sci.*, 2018, **4**, 1134–1145.
10. R.-R. Griffiths, *et al.*, Constrained Bayesian Optimisation for Automatic Chemical Design, *Chem. Sci.*, 2020, **11**, 577–586.
11. M. Parra-Prego, *et al.*, High-Throughput Experimentation Meets Machine Learning in Reaction Optimisation, *Angew. Chem., Int. Ed.*, 2021, **60**, 10756–10762.
12. N. S. Eyke, B. Koscher and K. F. Jensen, Toward Machine-Learning-Guided Experimental Design for Chemical Systems, *Trends Chem.*, 2020, **2**, 100–112.
13. B. J. Shields, J. Stevens and L. Cronin, Bayesian Reaction Optimisation as a Tool for Chemical Synthesis, *Nat. Rev. Chem.*, 2021, **5**, 315–330.
14. D. E. Fitzpatrick and S. V. Ley, Engineering Chemistry: Integration of Flow, Analytics and Control, *Org. Process Res. Dev.*, 2021, **25**, 1864–1884.
15. I. Rocha, *et al.*, Soft Sensors in Chemical and Bioprocesses: A Review, *Chem. Eng. Sci.*, 2020, **226**, 115832.
16. A. M. Schweidtmann, *et al.*, Deep Reinforcement Learning in Chemical Reaction Optimisation and Synthesis Planning, *AIChE J.*, 2019, **65**, e16690.
17. P. Schwaller, *et al.*, Predicting retrosynthetic disconnections and reaction conditions with transformer-based models, *Mach. Learn.: Sci. Technol.*, 2021, **2**, 015016.
18. C. W. Coley, *et al.*, Autonomous Discovery in the Chemical Sciences Part I: Progress, *Angew. Chem., Int. Ed.*, 2022, **61**, e202204245.
19. D. C. Elton, Z. Boukouvalas, M. D. Fuge and P. W. Chung, Deep Learning for Molecular Design—A Review, *J. Chem. Inf. Model.*, 2019, **59**, 1096–1108.
20. W. H. Green, *et al.*, Data and Uncertainty in Chemical Kinetics and Catalysis, *ACS Catal.*, 2021, **11**, 10672–10696.

3 Revolutionising Drug Discovery and Development

3.1 Chapter Overview

Drug discovery has always been a game of winnowing: from millions of possibilities to a handful of candidates that can be made into robust pharmaceutical products and dosage forms, are safe and effective, and can be delivered to patients. Over the past decade—and decisively in the last two years—artificial intelligence (AI) has moved from the periphery of this enterprise to its centre. Rather than replacing scientific judgement, AI has become a force multiplier for chemists and drug developers: accelerating hypothesis generation, sharpening experiment planning, and focusing scarce laboratory effort on the most informative tests.

This chapter explains, in a chemistry-first way, how AI is transforming the pharmaceutical pipeline:[2] from target identification and validation (TID/TV), through hit finding and lead optimisation (LO), to preclinical ADME/Tox, formulation and dosage-form development, clinical trial design, and post-marketing safety. We differentiate hype from practice[3] by grounding each section in executable workflows and peer-reviewed evidence—highlighting where AI delivers value today, and where caution, governance and rigorous validation remain non-negotiable. You will see the rise of graph neural networks (GNNs) for property prediction, the shift from brute-force virtual screening to active learning and generative design, the coupling of retrosynthesis planning with autonomous platforms to close the loop from *in silico*

RSC Foundations No. 7
AI Revolution in Chemistry
By Brian McKew
© Brian McKew 2026
Published by the Royal Society of Chemistry, www.rsc.org

ideas to wet-lab proof, and the introduction of AlphaFold 3 (AF3) into structure-enabled discovery. We also examine how AI supports clinical development—from patient stratification to operational forecasting—within the constraints of GxP and regulatory expectations.

By the end, you will have a detailed map of how to assemble AI-enabled workflows for your programmes: which data to curate, which models to consider, how to structure validation, where to bring in automation, and how to integrate makeability, CMC, pharmaceutical formulations and processes, and safety so that AI gains persist beyond a single case study.

3.2 Core Concepts and Definitions

3.2.1 From Informatics to AI-native Discovery

Cheminformatics provided descriptors, similarity search, QSAR, docking and pharmacophores. AI augments these with models that learn from examples—supervised, self-supervised or reinforcement learning—to predict molecular properties, biological activity and clinical signals, or to generate molecules that meet multiple constraints. Modern systems assimilate heterogeneous data (structures, images, omics, text, clinical variables) and optimise to explicit task-level metrics (*e.g.*, ROC-AUC, mean absolute error, expected improvement) (Figure 3.1).

3.2.2 The Funnel Reframed as Decisions with Uncertainty

TID/TV. Inputs: multi-omics, literature, phenotypic screens, pathway maps. Decision: advance a target and choose modality (small molecule, PROTAC, antibody, RNA). AI: causal inference, network biology, text mining.

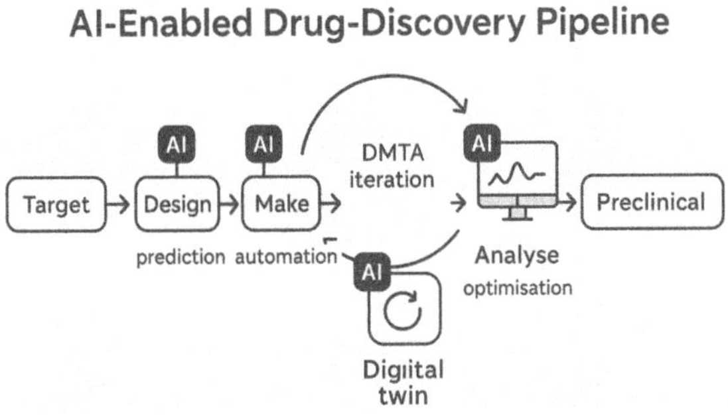

Figure 3.1 AI-enabled drug-discovery pipeline.

Hit ID. Inputs: vendor/in-house/virtual libraries,[11] target info (sequence/structure), prior SAR, assay design. Decision: which compounds to test next. AI: ligand-based *vs.* (GNNs/transformers), structure-based re-scoring, generative seeding.

LO. Inputs: hits, medicinal chemistry hypotheses, ADME/Tox assays, synthetic routes. Decision: which analogues to make next. AI: multi-objective optimisation (potency, selectivity, solubility, permeability, metabolic stability, hERG, CYP, clearance), active learning, retrosynthesis feasibility.

Preclinical. Inputs: *in vitro* ADME/Tox, *in vivo* PK/PD, safety pharmacology, CMC. Decision: nominate development candidate; plan scale-up. AI: ADME/Tox prediction, formulation support, process modelling (digital twins, soft sensors).

Clinical. Inputs: epidemiology, RWE, biomarker strategies, exposure–response models. Decisions: dose, regimen, endpoints, sample size, adaptive rules. AI: patient/site selection, screening failure prediction, enrolment forecasting, signal detection (with governance).

3.2.3 Data and Representations—What Models See

Molecules. SMILES (canonical, standardised) for sequence models; graphs (atoms/bonds as nodes/edges) for GNNs;[7] 3D conformers for geometry-dependent properties (pK_a, $\log D$, solvation, permeability) and for docking/physics-based rescoring. Apply tautomer normalisation, stereochemistry handling, charge normalisation, and salt/solvent stripping.

Proteins and complexes. Targets as sequences (for language models), 3D structures (X-ray/cryo-EM/NMR/AF-class), and binding-site graphs/grids. Assemble off-target panels and multi-task models for polypharmacology.

Assays. Preserve raw and processed data with metadata:[9] cell line, passage, media, temperature, pH, plate layout, instrument settings, internal standards, QC flags (*e.g.*, Z'), and batch effects. For phenotypes, retain images/time-series for representation learning.

Clinical. Harmonise features, comorbidities, concomitants, omics, endpoints, longitudinal outcomes; encode as tabular/time-series; respect privacy and governance.[19]

3.2.4 Model Families You Will Actually Use

Supervised learners (random forests, gradient-boosted trees, GNNs, transformers) for property/activity prediction.

Self-supervised molecular/protein LMs for chemically meaningful embeddings; support few-shot tasks.

Generative models (variational autoencoders, diffusion, RL) for *de novo* design under constraints (potency windows, ADME/Tox thresholds, physicochemical limits, synthesise-ability).

Active learning and BO for experiment selection.

Causal inference (propensity scores, instrumental variables, TMLE) for target and effect estimation.

3.3 Methods and Workflows

3.3.1 Target Identification and Mechanism Hypotheses

Data fusion and knowledge graphs. Build a target–pathway–phenotype graph integrating GWAS, expression/proteomics, perturbation signatures and curated pathway maps. Use GNNs over heterogeneous networks to score gene–disease links; apply NLP to extract mechanistic assertions from publications/patents. Prioritise targets with convergent evidence and tractability.

AF3-enabled structural hypotheses. Where structures are missing, AlphaFold 3 predicts protein–ligand and protein–nucleic acid complexes.[1] Use confidence metrics and consider alternative conformers; treat AF3 as hypothesis generation for docking/FEP, mutagenesis and biophysics—not as ground truth.

3.3.2 Checklist—TID/TV

Target dossier (function, disease linkage, tractability, safety red flags).

Knowledge graph with genetic and literature evidence; rank by strength and network centrality.

AF-scale mapping of ligandable pockets and family-selectivity liabilities.

Validation plan (CRISPR KO/knockdown, probes, phenotypic rescue).

Go/no-go criteria (effect size, translatability markers, safety signals).

3.3.3 Hit Identification—Model-guided Triage

Modern virtual screening. Classical docking scales but suffers from scoring bias. Improve early enrichment by coupling fast docking to learned re-scorers (3D-CNN/graph scorers trained on curated complexes). Ligand-based models (GNNs/transformers) triage vendor space before docking.

Active learning (AL) library design. Start with a diverse seed screen (1–5 k compounds), train a classifier/regressor, and iteratively select high-value batches by uncertainty/diversity criteria. Expect 2–3× enrichment at fixed budget and discovery of non-obvious scaffolds beyond nearest neighbours.

Generative seeding.[6] Use conditional diffusion (or RL) to propose novel yet synthesizable seeds. Enforce physicochemical windows (300–450 Da, $c \operatorname{Log} P$ 1–3, TPSA 60–90 Å^2), remove alerts, and require retrosynthesis feasibility (SCScore/RAscore ≤ threshold, ≤ 4–6 steps, allowed reaction classes). Export candidates to CASP and synthesis.

Assay design and analysis. Quantify Z'/dynamic range; include orthogonal readouts (*e.g.*, thermal shift + enzymatic turnover). For phenotypes, use imaging + self-supervised vision to cluster morphologies and connect to gene signatures.

Follow-up. Confirm hits across orthogonal assays; filter PAINS/ aggregators; run triage ADME (solubility, permeability, microsomal stability, hERG/CYP panels) to avoid non-developable matter.

3.3.4 Lead Optimisation—Multi-objective Design in Closed Loop

Define the multi-objective (MPO). Combine potency with ADME/Tox and makeability: solubility, permeability (P_app/log D7.4), microsomal stability (HLM-CL_int), efflux (*e.g.*, P-gp), hERG, CYP, protein binding,[13] synthetic tractability (steps/route risk). Use desirability functions or Pareto fronts to align chemists and models (Figures 3.2 and 3.3).

A schematic overview illustrating an iterative cycle in which AlphaFold-predicted protein pocket geometries inform analogue

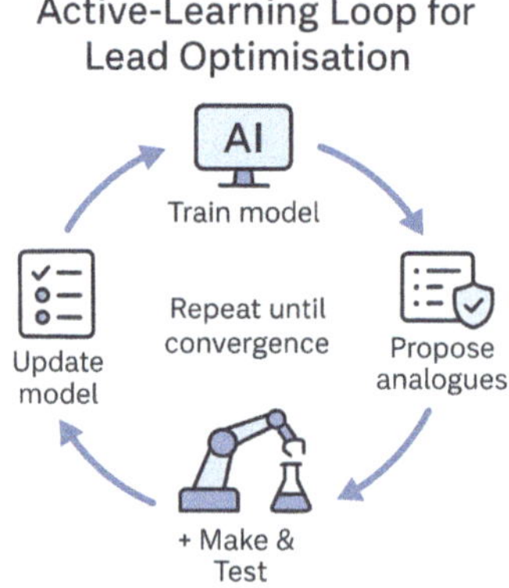

Figure 3.2 Active-learning loop for lead optimisation.

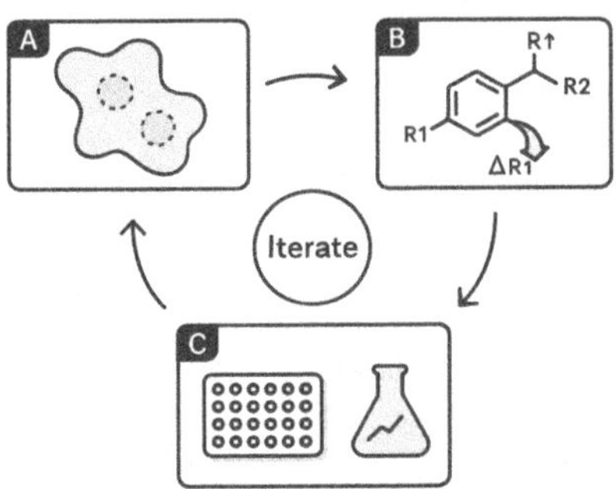

Figure 3.3 AlphaFold-guided analogue design loop.

modification and testing. Panel A (top left) depicts an abstract contour representation of a protein binding pocket with dotted "hotspot" regions. Panel B (top right) shows a generic ligand scaffold with variable groups R_1 and R_2 and curved arrows ΔR_1 and ΔR_2 indicating substitution exploration. Panel C (bottom centre) combines a simplified assay microplate grid and a small line chart representing phenotypic readout. Curved arrows link $A \rightarrow B \rightarrow C \rightarrow A$ around a central hub labelled "Iterate", symbolising design–test–refine feedback.

Property models with GNNs/transformers.[8]

Splits: Use Bemis–Murcko scaffolds or time-splits[14] to measure true generalisation.

UQ: Apply conformal prediction/ensembles; set decision thresholds linked to precision/recall.

Explainability: Atom-level attributions support SAR debates (*e.g.,* heteroatom contributions to hERG risk).

AL/BO for analogue selection.

Acquisition: Thompson sampling/UCB with diversity constraints.

Feasibility: Enforce retrosynthesis filters (allowed transformations, step caps, inventory), bias towards platform chemistry; prioritise molecules compatible with current routes.

Batching: 24–48 molecules per sprint, with 10–20% reserved for exploration outside current chemotypes.

Generative co-pilot.

Constraints first: Forbid reactive moieties/toxophores; bind generation to fragment merges that the lab can make; condition on physicochemistry.

Scoring: Include potency/selectivity predictors, ADME/Tox surrogates, and route risk (SCScore) in a single multi-objective score.

Sanity checks: Drop any design >6 steps or with low-confidence disconnections; keep a small wildcard slice to sample "unknown-unknowns".

DMTA cadence.

Design–Make–Test–Analyse on short cycles: CASP emits actionable procedures[10,15] (χDL/robot-ready steps); miniaturised parallel synthesis/flow[12] with automated work-up; at-line analytics (UPLC-UV/MS).

Condition BO runs in parallel to lift yields/selectivities; capture robustness with replicate/perturbation tests.

Prospective validation. Each iteration commits to making and testing a representative set (*e.g.*, 16–48 analogues) with pre-declared success criteria (Δ-potency, ADME gains, novelty, synthetic yield). Track predicted *vs.* observed; recalibrate; update acquisition and goals.

3.3.5 Preclinical Development—ADME/Tox and Developability

In silico ADME/Tox: Multi-task learners (GNNs/transformers) triage permeability, solubility, clearance, CYP inhibition/induction, hERG, DILI, Ames. Apply assay-specific models where possible; compute AD and decline out-of-domain predictions; prioritise experiments for uncertain regions.

Translational modelling: Pair ML-based human PK (clearance, volume, F) with PBPK or simplified PK; explore dose–exposure–response spaces and sensitivities (solubility *vs.* permeability *vs.* clearance). Tailor descriptors for modalities (macrocycles/PROTACs: 3D PSA, Fsp^3, chameleonicity; permeability mechanisms).

CMC/process foresight: Use yield/risk predictors and route feasibility to avoid dead-ends; collaborate early with process chemists; employ digital twins/soft sensors to de-risk solvent swaps, crystallisation, and scale-out before formal development.

3.3.6 Clinical Development—Where AI Helps (and Where It Cannot Replace Evidence)

Patient stratification: Supervised/unsupervised models identify genomic/phenotypic clusters for enrichment. Prefer transparent features (*e.g.*, composite scores, validated biomarkers); pre-specify hypotheses; audit for bias.

Trial design and operations:[18] AI supports site selection, screen-failure prediction, enrolment forecasting; NLP accelerates protocol authoring, eligibility checking, and safety narrative drafting. For digital endpoints, validate feature extraction pipelines and device calibration; monitor signal quality continuously.

Adaptive/Bayesian designs: Bayesian frameworks formalise interim decisions (arm dropping, sample-size re-estimation) under error control. AI can simulate many synthetic trials to assess power and operating characteristics, but statistical governance is essential.

Pharmacovigilance and label expansion. Post-marketing, ML triages spontaneous reports/EHRs and literature for potential signals (disproportionality, NLP case-finding). Causality remains clinical; AI surfaces patterns for expedited human review.

3.4 Case Studies and Recent Signals (2024–2025)

3.4.1 Case 1—An AI-designed Small Molecule with Phase 2a Signal (2025)[5]

A *de novo*–designed TNIK inhibitor for idiopathic pulmonary fibrosis (often cited as rentosertib/INS018_055) reported randomised, placebo-controlled Phase 2a results: acceptable safety/tolerability and dose-dependent improvements in forced vital capacity over 12 weeks. While confirmatory trials are required for efficacy/safety, this peer-reviewed signal validates the plausibility of end-to-end AI[20] from target to early human readouts. Lessons: enforce synthesise-ability and ADME/Tox constraints in design; require prospective validation at each stage; and run CMC in lockstep with chemistry to avoid late route surprises.

3.4.2 Case 2—ML-assisted Antibiotic Discovery Against *A. baumannii*[4]

A DL pipeline trained on antimicrobial datasets triaged millions of molecules to a compact set of novel scaffolds with potent activity, culminating in abaucin, a selective inhibitor of Acinetobacter baumannii with *in vivo* efficacy in murine models. The replicable template: careful phenotypic assay design, large-scale representation learning, mechanistic follow-up (target deconvolution, resistance mapping), and prioritisation for medicinal chemistry feasibility.

3.4.3 Case 3—Closed-loop Medicinal Chemistry and Reaction Optimisation (2024)

Bayesian optimisation combined with automated synthesis/analysis[17] achieved >80% yields while sampling <5% of candidate spaces; allied principles in medicinal chemistry (MPO + AL + retrosynthesis filters + mini-flow synthesis) shorten DMTA cycles and raise learning per experiment.

3.4.4 Case 4—Structure-guided Design in the AF3 Era

With AF3 delivering complex models, teams report accelerated design sprints against previously intractable targets: refined binding hypotheses, rational scaffold morphing, and focused synthesis of conformation-stabilising analogues. Not universal—pockets may remain dynamic or mis-modelled—but the cost of exploring structural ideas has dropped sharply, encouraging combined use of physics-based methods (*e.g.*, FEP) and structure-guided generative design.

3.5 Common Pitfalls and How to Mitigate Them

3.5.1 Data Leakage and Unrealistic Validation

Risk: Random splits inflate SAR performance; models "learn the assay" (plate/batch) rather than chemistry.

Mitigation: Scaffold or time-split validation; hold operational IDs for diagnostics only; prospective tests; monitor drift as chemistry evolves.

3.5.2 Out-of-domain Deployment

Risk: Applying kinase-trained models to ion-channel ligands; deploying hERG model beyond training chemical space.

Mitigation: Compute applicability domain; enforce confidence thresholds; extend training *via* active learning; publish model cards.

3.5.3 Optimising Proxies, Not Outcomes

Risk: Chasing docking scores/surrogates that do not track efficacy/safety; ignoring synthesis/CMC constraints.

Mitigation: Design multi-objective MPO aligned to project goals; require orthogonal assays; embed route risk in scoring.

3.5.4 Synthesis Bottlenecks

Risk: Generative designs that cannot be made at an acceptable yield/scale.

Mitigation: Enforce retrosynthesis filters (SCScore/RAscore; step caps; allowed reactions); co-optimise reaction conditions.

3.5.5 Black-box Opacity

Risk: Models influence nomination without transparency or challenge.

Mitigation: Use attributions/counterfactuals; benchmark against baselines; require governance (documentation, change control, audit trails).

3.5.6 Over-extrapolating Early Clinical Signals

Risk: Reading too much into small, short trials; assuming AI involvement reduces attrition later.

Mitigation: Treat early readouts as hypothesis-generating; power confirmatory trials; use external controls with safeguards only.

3.6 Implementation Guidance (Step-by-step)

3.6.1 Launch an AI-enabled Hit-finding Campaign

Frame the decision—*e.g.*, "From 2 million supplier compounds, triage 10 000 to maximise true hit rate at 10 µM single-point".

Curate data—normalise units; remove salts/duplicates; define positives/negatives; detect confounders (plate effects, autofluorescence).

Seed screen—run a diverse 1–5k set; compute Z' and baseline hit rate; assemble a labelled training set with explicit negatives.

Train models—baseline (RF/XGBoost) *vs.* GNN/transformer; evaluate with scaffold/time splits; report ROC-AUC, PR-AUC, top-k enrichment.

Active learning—pick batches *via* uncertainty + diversity; cap per-chemotype density; allocate a random exploration fraction.

Assay hygiene—add orthogonal assays; track PAINS/aggregators; estimate false-positive rates.

Make *vs.* buy *vs.* generate—generative seeding under constraints; validate *via* retrosynthesis; split budget responsibly.

Decision gates—predefine potency/selectivity/solubility/liability thresholds and fail-fast rules.

3.6.2 Close the LO Loop (Design → Make → Test → Analyse)

Define MPO—*e.g.*, maximise pIC_{50}; minimise hERG and CL_int; keep $c \operatorname{Log} P$ 1–3, TPSA 50–90 Å^2; SA ≤ 5; route ≤ 5 steps.

Build/update property models—GNN/transformer; adopt conformal prediction; monitor calibration.

Propose analogues—combine R-group enumerations and generative proposals under constraints.

Filter by makeability[16]—enforce allowed reactions, step caps, and inventory; exploit platform chemistry.

Prioritise—Thompson/UCB with diversity; 24–48 molecules per sprint; 10–20% exploration.

Synthesis and analytics—HTE/flow; automated work-up; standardised LC–MS/UPLC; batch QC.

Prospective validation—compare predicted *vs.* observed; update models; maintain model cards.

Iterate—adjust MPO weights as biology/CMC evolve; maintain a backlog for continuity.

3.6.3 Integrate with Preclinical and Clinical Partners

Align early with DMPK/toxicology—co-design prediction panels; agree go/no-go thresholds.

Translational modelling—share exposure predictions/risk flags; co-develop dose-finding simulations; validate assumptions.

Trial ops—small, transparent models for screen-failure/site performance; bias checks; robust data monitoring.

3.7 Key Takeaways

AI complements medicinal chemistry—use models as decision aids, corroborated by orthogonal evidence and prospective testing.

Define decisions and success criteria up front; align models and experiments to project-level objectives (MPO + makeability + cost).

Close the loop with active learning/BO; embed retrosynthesis and green/safety constraints in selection.

Exploit structural AI judiciously—AF3 hypotheses guide, not replace, experiments; validate by docking/FEP, mutagenesis, biophysics.

Front-load ADME/Tox and CMC—multi-task predictors, soft sensors and twins reduce late attrition.

Governance matters—scaffold/time splits, AD/UQ, model cards, change control and drift monitoring.

Clinical decisions demand transparency—AI sharpens design/operations but does not substitute for clinical evidence.

Data foundations—clean, richly annotated data and metadata are the strongest predictors of AI ROI.

References

1. J. Jumper, R. Evans and A. Pritzel, *et al.*, Highly accurate protein structure prediction with AlphaFold 3, *Nature*, 2024, **627**, 795–803.
2. J. Vamathevan, D. Clark and P. Czodrowski, *et al.*, Applications of machine learning in drug discovery and development, *Nat. Rev. Drug Discovery*, 2019, **18**, 463–477.
3. A. Bender and M. Cortés-Ciriano, Artificial intelligence in drug discovery: what is realistic, and what are the pitfalls?, *Drug Discovery Today*, 2021, **26**(2), 511–524.
4. J. M. Stokes, K. Yang and K. Swanson, *et al.*, A deep-learning approach to antibiotic discovery, *Cell*, 2020, **180**(4), 688–702.e13.
5. A. Zhavoronkov, Y. A. Ivanenkov and A. Aliper, *et al.*, Deep learning enables rapid identification of potent DDR1 kinase inhibitors, *Nat. Biotechnol.*, 2019, **37**, 1038–1040.
6. W. P. Walters, R. Barzilay and M. Murcko, Applications of deep learning in molecule generation and property prediction, *ACS Cent. Sci.*, 2020, **6**(10), 1722–1731.
7. Z. Wu, B. Ramsundar and E. N. Feinberg, *et al.*, MoleculeNet: a benchmark for molecular machine learning, *Chem. Sci.*, 2018, **9**, 513–530.
8. K. Yang, K. Swanson and W. Jin, *et al.*, Analyzing learned molecular representations for property prediction, *J. Chem. Inf. Model.*, 2019, **59**(8), 3370–3388.
9. K. Huang, T. Fu and W. Gao, *et al.*, Therapeutics Data Commons: ML datasets and tasks for drug discovery, *Sci. Data*, 2021, **8**, 316.
10. P. Schwaller, T. Laino and T. Gaudin, *et al.*, Molecular transformer: uncertainty-calibrated reaction prediction, *ACS Cent. Sci.*, 2019, **5**(9), 1572–1583.
11. J. J. Irwin, K. G. Tang and J. Young, *et al.*, ZINC20—ultra-large chemical database for ligand discovery, *J. Chem. Inf. Model.*, 2020, **60**(12), 6065–6073.
12. B. P. MacLeod, F. G. Parlane and T. D. Morrissey, *et al.*, Self-driving laboratory for thin-film materials, *Sci. Adv.*, 2020, **6**, eaaz8867.
13. H. Li, C. W. Yap and C.-Y. Ung, *et al.*, ML approach for predicting hERG inhibition, *Bioinformatics*, 2021, **37**(24), 4710–4718.
14. R. P. Sheridan, Time-splits as validation for QSAR models, *J. Chem. Inf. Model.*, 2013, **53**(4), 783–790.
15. M. H. S. Segler, M. Preuß and M. P. Waller, Planning chemical syntheses with DNNs and symbolic AI, *Nature*, 2018, **555**, 604–610.
16. C. W. Coley, W. H. Green and K. F. Jensen, Machine learning in CASP, *Acc. Chem. Res.*, 2018, **51**(5), 1281–1289.
17. W. Gao, C. Do and Y. Tian, Sample-efficient deep molecular reaction optimisation *via* RL, *Nat. Mach. Intell.*, 2022, **4**, 441–451.
18. S. Kannan, *et al.*, Artificial intelligence for clinical trial design, *Trends Pharmacol. Sci.*, 2022, **43**(9), 730–746.
19. V. Kehl and B. Ulm, *et al.*, Federated learning in pharmaceutical research, *NPJ Digital Med.*, 2023, **6**, 88.
20. S. Ekins, *et al.*, Exploiting machine learning for end-to-end drug discovery and development, *Nat. Mater.*, 2023, **22**, 1–15.

4 AI-enabled Materials Discovery and Optimisation

4.1 Chapter Overview

Materials science asks a deceptively simple question with daunting breadth: which arrangement of atoms, processed in which way, will manifest the property we need—reliably, at scale, and with acceptable cost and sustainability? For most of the twentieth century the answer came from painstaking theory-informed exploration and intuition honed over years in the lab. In the twenty-first, artificial intelligence (AI) adds a second engine to that endeavour—one that can scan vast compositional and structural spaces, learn patterns from heterogeneous data (computations, measurements, images, text), propose candidates and processing windows, and close the loop with automation to test, learn, and improve.

This chapter explains how AI accelerates the full design–make–measure–learn cycle for materials[8] (Figure 4.1). We begin by introducing the key representations (composition-only features, crystal graphs, microstructure images), learning regimes (supervised, self-/physics-informed, multi-fidelity), and generative approaches (inverse design *via* diffusion and related models). We then build practical workflows: curating and featurising data; training and validating property predictors; prioritising *via* active learning and Bayesian optimisation; interpreting stability and synthesise-ability; predicting processing–structure–property links; and integrating with self-driving

RSC Foundations No. 7
AI Revolution in Chemistry
By Brian McKew
© Brian McKew 2026
Published by the Royal Society of Chemistry, www.rsc.org

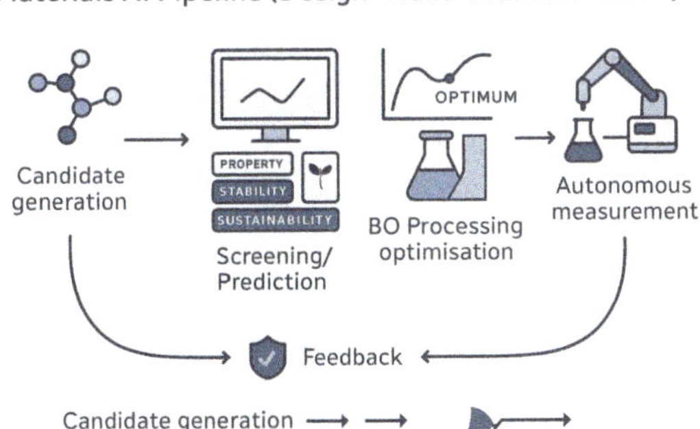

Figure 4.1 End-to-end AI pipeline for materials (design–make–measure–learn).

laboratories (SDLs). Throughout, we show how to embed sustainability—criticality, embodied energy, toxicity, and recyclability—as first-class design constraints, not after-the-fact filters.

Recent results make the opportunity plain. Graph and diffusion models have proposed hundreds of thousands of stable crystals;[6] autonomous labs have validated dozens within days to weeks; Bayesian optimisation has pushed device-relevant metrics while exploring only a few per cent of the combinatorial space. Polymer design teams have used ML to steer degradability without giving up strength; thin-film discovery groups have turned micro-fabrication lines into closed-loop platforms. AI does not replace physics, processing knowledge, or the craft of experimentation—but it does focus all three, turning "what might work" into "what we should try next" with unprecedented efficiency.

4.2 Core Concepts and Definitions

4.2.1 The Materials Design Problem, Rephrased

The canonical materials question is inverse: given a target property vector P (*e.g.*, ionic conductivity, band gap, modulus, thermal stability), find a structure–composition–process tuple (S, C, Π) that realises it under cost/sustainability constraints. AI addresses all three levers:

Composition (C): which elements and stoichiometries?

Structure (S): which crystal symmetry, lattice parameters, local environments, defect chemistry?

Process (Π): which synthesis route, temperatures/pressures/ atmospheres/solvents/anneals, thin-film deposition recipe, or microstructure control?

The right AI methods depend on the information you can encode about these levers and the amount/quality of data you have.

4.2.2 Representations: What the Model "Sees"

Composition-only features[25] (when no structures are available): statistics over elemental properties (electronegativity, atomic radius, valence electron count), Magpie/matminer descriptors, learned composition embeddings. Useful for early screening, polymers, glasses, or when multiple structural prototypes exist.

Crystal graphs for periodic solids: atoms as nodes; bonds/near-neighbour relations as edges; periodic boundary conditions. Variants include CGCNN,[1] MEGNet,[2,17] ALIGNN; edges can encode distances, bond orders, angular information, or symmetry-aware features (Figure 4.2).

3D grids/point clouds: voxelised electron densities or atomic potentials; helpful for learned scoring functions and microstructure-sensitive properties.

Microstructure images and spectra:[11] SEM/TEM micrographs, XRD patterns, Raman/IR spectra. Convolutional/transformer vision backbones learn microstructural descriptors that correlate with mechanical or transport properties and that support quality control.

Text/literature: pretrained language models (*e.g.*, mat2vec-style)[4] embed synthesis recipes and outcomes; extraction tools convert prose ("calcined at 750 °C under N_2 for 4 h") into actionable process parameters.

4.2.3 Learning Regimes You Will Actually Use

Supervised prediction: property regressors/classifiers trained on DFT or experimental labels (formation energy, band gap, elastic constants,

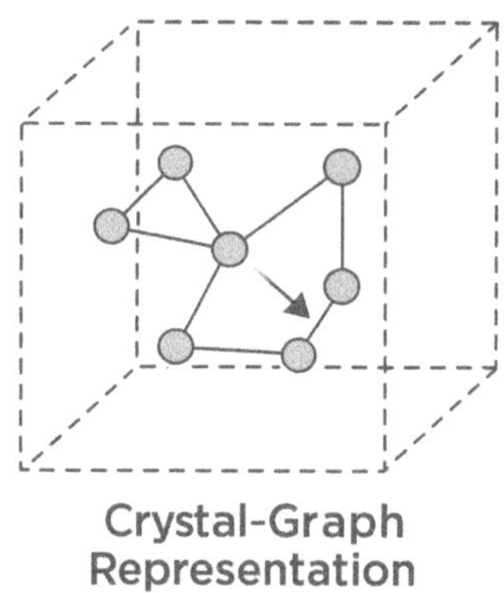

Figure 4.2 Crystal-graph representation for periodic solids.

ionic conductivity, thermal stability, strength).[15] Use when labels are plentiful enough to generalise.

Self-supervised/representation learning: pretrain on unlabelled structures or literature to learn chemically meaningful embeddings; fine-tune on small labelled sets (few-shot regimes).

Physics-informed and multi-fidelity: combine low-fidelity (cheap) and high-fidelity (expensive) sources, calibrating one to the other; inject constraints (symmetry, stoichiometry, charge balance, positive-definite tensors) to prevent unphysical predictions.

Active learning (AL) and Bayesian optimisation (BO): pick what to compute or measure next based on uncertainty and potential for improvement. This is critical when DFT/experiments are costly.

Generative (inverse design): diffusion/flow/VAEs that propose valid crystals or polymers, optionally conditioned on desired properties and constraints (*e.g.*, space group, composition ranges, synthesise-ability).

4.2.4 Stability, Synthesise-ability, and Process Realism

AI can suggest structures anywhere in the chemically permissible search space; the art is narrowing to what you can actually make and keep:

Thermodynamic plausibility: predicted formation energy, decomposition energy, and (for solids) phonon stability; position relative to the convex hull of competing phases.

Kinetic accessibility and defect chemistry: synthesis-route priors from literature mining; preferred oxidation states; diffusion pathways; defect formation energies when available.

Processing windows: temperature–pressure–atmosphere ranges, precursors and solvents, anneal schedules, and thin-film deposition parameters; volatility/toxicity risk.

Lifecycle and sustainability: presence of critical raw materials (CRMs), embodied energy and carbon, toxicity, solvent impact, recyclability, and supply risk.

Treat these not as after-the-fact filters, but as constraints integrated with property objectives from the outset.

4.3 Methods and Workflows

4.3.1 Data Curation and Labelling

Source integration: Combine public computational repositories (Materials Project,[9] OQMD, JARVIS[10]) with in-house DFT and experiments (powders, thin films, polymers). For experiments, require units and uncertainty; record substrate, thickness, atmosphere, humidity, temperature history, and instrument calibration. For computations, record functionals, cut-offs, k-meshes, smearing, and relaxation criteria.

De-duplication and equivalence: Canonicalise composition (reduced formula), symmetry, and cell settings; cluster near-duplicate structures (prototype twinning, trivial relabellings). Align temperatures/pressures for cross-study comparability.

Label hygiene: For DFT labels with systematic bias (*e.g.*, band gaps), either post-calibrate (learned correction) or restrict training to homogeneous label sets. For noisy properties (*e.g.*, thin-film mobility), favour distributional targets or replicate means with variance.

Splits to match deployment. Avoid over-optimistic performance by using time-based (train $<$ year t, test $\geq t$), structure-wise (no shared prototypes), or composition-wise splits. Keep an external test set untouched until the end.

4.3.2 Feature Engineering and Representation

Composition-only baselines:[5] Start with matminer/Magpie descriptors or learned composition embeddings. They often explain substantial variance and help triage which property is learnable with available data.

Crystal graphs: For solids with known structures, train CGCNN/ MEGNet/ALIGNN-style models. Encoding choices matter: neighbour counts; distance/angle bins; long-range interactions; symmetry-aware pooling. For microstructure-sensitive properties, consider dual-branch models that ingest crystal graphs and image/spectral features.

Literature-derived process priors: Use text mining to build distributions for synthesis parameters by composition family (*e.g.*, "perovskite oxides: 800–1100 $^\circ$C in O_2 for 2–10 h"), which later serve as priors for BO of processing.

4.3.3 Training, Validation, Uncertainty, Domain

Baselines first: Linear/tree models with composition features are invaluable sanity checks; they surface pathologies (*e.g.*, properties linked to atomic number proxies).

Advanced models next: Fit GNNs/transformers when the task warrants structure sensitivity. Log and compare MAE/RMSE[3] (regression) and ROC-AUC/PR-AUC (classification) across structure-wise and time-wise splits. Track calibration with reliability diagrams.

Uncertainty and applicability domain (AD):[20] Use deep ensembles, Monte-Carlo dropout, or evidential regression for uncertainty; compute AD *via* distance in learned embedding space. Treat AD breaches as "I don't know"; route them to AL or human review.

Robust metrics: Report performance distributions (median/percentiles) and ablations (*e.g.*, removing elemental groups) to expose shortcut-learning. For device-level surrogates (*e.g.*, perovskite solar-cell efficiency), use multi-task models to control for publication bias.

4.3.4 Screening, Prioritisation, and Active Learning

Two-stage triage: Stage 1: cheap composition-only models eliminate obvious non-candidates; Stage 2: expensive structure-aware models rank survivors. Combine with constraints (CRMs, toxicity, thermal budgets).

AL with batch diversity:[16] Use AL to pick what to compute next (DFT or surrogate) and what to synthesise. Acquisition functions[14] (expected improvement, UCB, expected hypervolume improvement for multi-objective) select promising points; diversity penalties prevent redundant batches.

BO for processing: Treat process parameters (temperatures/flows/atmospheres/anneals) as the design space; couple BO to inline metrics (conductivity, PL, XRD peak features). Integrate feasibility models to avoid unsafe or implausible regimes.

4.3.5 Generative (Inverse) Design

Crystal generation:[7] Diffusion/flow models[21] can sample plausible structures under composition and space-group constraints; property-conditioned variants bias toward target windows (*e.g.*, band gap $\sim$1.6 eV; wide phonon gap). For each sample: (i) symmetry clean-up, (ii) DFT relaxation, (iii) stability checks (formation/decomposition energy; phonons if feasible), and (iv) screen with property predictors.

Polymer and soft-matter design: Sequence- or graph-based generators propose monomers and architectures guided by property predictors (T_g, modulus, gas permeability) and degradability surrogates. Add synthesis-route and processability constraints (solubility windows, solvent selection, toxicity).

Multi-objective and constrained generation: Build composite scores balancing performance with sustainability (criticality, embodied energy/carbon, hazardous solvent use) and cost. Penalise CRMs, halogenated solvents, PFAS-like moieties; prioritise elements with robust supply chains.

4.3.6 Predicting Synthesise-ability and Processing Windows

Thermodynamics plus "soft" priors: Predict formation energy, then query competing phases; estimate metastability windows; combine with literature-derived priors for temperature/atmosphere/time by composition family. For thin films, learn composition–process–phase maps (*e.g.*, precursor ratios → phase fractions).

Text-to-process models: Use literature extraction to propose initial synthesis conditions; then tighten with BO guided by inline XRD/PL/Raman/ellipsometry: Capture failure modes (clogging, delamination, volatility) for feasibility models.[24]

4.3.7 Closing the Loop with Self-driving Labs (SDLs)[13]

4.3.7.1 Minimal SDL for Materials

Planner: AL/BO service proposing next compositions/conditions.

Deposition/synthesis: Spin/spray/inkjet coaters; sputter/ALD; sol–gel; combustion synthesis; powder solid-state routes; hydro/solvo-thermal autoclaves.

Analytics: Inline/at-line XRD; PL/UV–vis; Raman/IR; 4-point probe; profilometry/ellipsometry; contact angle.

Orchestration: Queue scheduler; recipe templates; data broker to assemble (composition, process, spectra, properties); state machine for safe halts.

Governance: Calibration SOPs; drift monitoring; interlocks; physical E-stops.

4.3.7.2 Multi-objective Operation

Use EHVI in BO to trade performance (*e.g.*, power factor, mobility) against cost/criticality or process risk; maintain a Pareto front for decision-makers (Figure 4.3).

4.4 Case Studies and Recent Signals (2024–2025)

4.4.1 Scaled Candidate Generation with Rapid Experimental Validation

Large-scale graph/diffusion pipelines have proposed $\sim 10^5$–10^6 plausible crystals, filtering by predicted stability and property windows

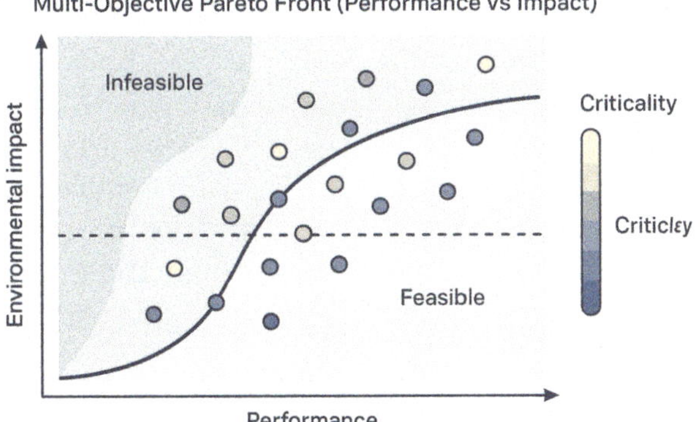

Figure 4.3 Multi-objective Pareto front (performance *vs.* embodied carbon *vs.* criticality).

(*e.g.*, band gap, elastic moduli). Autonomous "A-Lab"-style platforms then synthesised and characterised dozens within days to weeks, validating that model → robot → measurement loops can shorten discovery cycles dramatically. The notable practice is not the raw number of candidates, but the front-loaded pruning (composition limits, space-group constraints, stability filters) that yields high hit density in the lab.

4.4.2   Multi-objective Optimisation in Thin-film Devices[12]

Perovskite and oxide-semiconductor groups have used BO with inline PL/UV–vis and four-point probe to optimise composition and anneals for target power factor/carrier mobility,[23] hitting near-optimal regimes after exploring ∼2–5% of nominal design space. Success factors: informative inline signals, careful humidity/temperature control, batch diversity in acquisitions, and robust post-hoc confirmation (independent device stacks, alternate substrates).

4.4.3   Polymer Discovery for Degradability[18] Without Giving Up Strength

Sequence- and graph-based models trained on curated polymer datasets predicted degradability proxies (hydrolysis rate classes) and mechanical properties. Generative models proposed monomers/ architectures inside regulatory solvent and toxicity envelopes. Small-batch synthesis validated that target degradability could be achieved while maintaining tensile strength/modulus within design bounds— an early but important signal for green polymer design.

4.4.4 Catalysts with Performance–Cost–Stability Trade-offs

Alloy/oxide catalyst teams used AL to navigate high-entropy compositional spaces under cost constraints. EHVI-driven BO identified compositions matching noble-metal performance with fractional precious-metal loadings, illustrating how cost and criticality can be integrated directly into the optimisation objective.

4.5   Common Pitfalls and How to Mitigate Them

4.5.1   Shortcut Learning and Data Leakage[22]

Risk: Models infer properties from dataset quirks (*e.g.*, elemental prevalence) rather than physics; near-duplicates leak across splits.

Mitigation: Use structure/time/composition-wise splits; ablate spurious features; report external test results; enforce training diversity.

4.5.2   Label Bias[22] and Fidelity Mismatch

Risk: Mixing DFT labels with heterogeneous settings; combining thin-film and bulk data; confusing computed and measured gaps.

Mitigation: Train on homogeneous labels or model multi-fidelity explicitly; post-calibrate; propagate metadata into models.

4.5.3   Ignoring Synthesise-ability and Processing

Risk: Generators propose metastable/impractical structures; predictors favour phases with vanishing synthesis windows.

Mitigation: Include stability/phonon checks; add process priors; use feasibility classifiers; penalise long/unsafe routes and volatile/toxic precursors.

4.5.4   Over-optimising Single Proxies

Risk: Maximising predicted band gap or a single figure of merit with poor device correlation; ignoring lifetime.

Mitigation: Use multi-objective BO; include stability/lifetime; confirm device-level performance for Pareto-front candidates.

4.5.5 Uncalibrated Uncertainty and Over-confident Deployment

Risk: Treating point predictions as certain; exploring only "best predicted" region.

Mitigation: Use ensembles/conformal/evidential methods; sample high-uncertainty points in AL; publish model cards with AD limits.

4.5.6 Sustainability as an Afterthought

Risk: High-performers rely on CRMs/hazardous solvents; routes with high embodied energy.

Mitigation: Add criticality/embodied-carbon/toxicity/cost to objectives/constraints; require green-chemistry metrics up front.

4.5.7 Underspecified Governance

Risk: No calibration logs; unversioned models; no safety interlocks in SDLs.

Mitigation: Adopt ALCOA+ practices; version data/models; implement change control; define safety envelopes and HAZOPs.

4.6 Implementation Guidance (Step-by-step)

4.6.1 Stand Up a Composition-to-property Screening Pipeline

Define the target (*e.g.*, wide-band-gap piezoelectric with $T_C > 450$ °C; solid electrolyte[19] with $\sigma_{ion} > 10^{-3}$ S cm^{-1} at 25 °C).

Assemble data from Materials Project/OQMD/JARVIS + in-house; unify units/functionals; time-stamp entries.

Featurise with matminer/Magpie; train baseline tree/linear models.

Escalate to crystal-graph models where structures exist; evaluate with time/structure splits; compute uncertainty and AD.

Filter by stability (decomposition/formation energies; phonons when feasible) and sustainability (CRM flags, embodied carbon).

Prioritise *via* AL to choose next DFT runs; retrain as labels accumulate.

4.6.2 Add Inverse Design and Synthesis Feasibility

Conditioned generation under composition/space-group constraints and targeted property windows.

Triage: Symmetry clean-up $\rightarrow$ DFT relaxation $\rightarrow$ stability checks $\rightarrow$ property predictors.

Feasibility: Literature-derived priors for T/p/atmosphere/time; penalise unsafe/long procedures; rank with route-risk scores.

Batch: Select diverse candidates with AL/BO balancing exploration/exploitation.

4.6.3 Optimise Processing *via* BO (Thin Films or Powders)

Design space: Temperatures/anneals, atmospheres, precursor ratios, substrate, solvent mix; set safety bounds.

Inline readouts: PL, UV–vis, Raman/IR, XRD peak features, 4-point probe, ellipsometry; enforce calibration SOPs.

BO loop: Surrogate + acquisition (EI/UCB/EHVI) propose 8–16-point batches; enforce diversity/feasibility; stop on convergence or budget.

Robustness: Re-measure best conditions across days/substrates; compute robustness bands.

4.6.4 Embed Sustainability by Design

Objective: Augment performance with criticality, embodied energy/carbon, toxicity, cost.

Constraints: Exclude PFAS-like motifs, highly toxic solvents, and CRMs where feasible; prefer reusable solvents and lower-temperature processing.

Decision artefacts: Deliver Pareto fronts and explainability (drivers of trade-offs) to stakeholders.

4.6.5 Minimal SDL for Materials (Starter Kit)

Hardware: Spin/spray/inkjet coater or powder-synthesis workstation; furnace/rapid thermal anneal; at-line XRD/PL/4-point probe; environmental control (RH/temperature).

Software: BO service; instrument APIs; queue-based orchestrator; ELN/LIMS integration.

Governance: Calibration SOPs; drift monitors; E-stops; change-control for models and recipes.

4.7   Key Takeaways

Representations drive learning: Composition features for broad triage; crystal graphs and images for structure-/microstructure-sensitive tasks.

Validation must match deployment: Use time/structure splits, uncertainty, and applicability domain; keep an external test.

Inverse design works when constrained: Bind generation to stability, synthesise-ability and sustainability from the outset.

Processing is a first-class variable: BO with informative inline readouts accelerates anneals/depositions/mixes under safety bounds.

Sustainability belongs inside the objective: Criticality, embodied carbon, toxicity, and cost should guide selection, not merely filter it.

SDLs are capabilities, not gadgets: Planner–executor–sensor–orchestrator stacks, calibration SOPs, and safety interlocks matter as much as models.

Speed: Large candidate sets + autonomous validation + BO deliver near-optima with few-percent exploration.

Governance underpins trust: ALCOA+ data, model cards, change control, and robust calibration are essential for durable impact.

References

1. T. Xie and J. C. Grossman, Crystal Graph Convolutional Neural Networks for an Accurate and Interpretable Prediction of Material Properties, *Phys. Rev. Lett.*, 2018, **120**, 145301.
2. C. Chen, W. Ye, Y. Zuo, C. Zheng and S. P. Ong, Graph Networks as a Universal Machine Learning Framework for Molecules and Crystals (MEGNet), *Chem. Mater.*, 2019, **31**, 3564–3572.
3. A. Dunn, Q. Wang, A. M. Ganose, D. Dopp and A. Jain, Benchmarking Materials Property Prediction Methods: Matbench, *npj Comput. Mater.*, 2020, **6**, 138.
4. V. Tshitoyan, *et al.*, Unsupervised Word Embeddings Capture Latent Knowledge from Materials Science Literature, *Nature*, 2019, **571**, 95–98.
5. L. Ward, *et al.*, A General-Purpose Machine Learning Framework for Predicting Properties of Inorganic Materials, *npj Comput. Mater.*, 2016, **2**, 16028.
6. A. Merchant, *et al.*, Scaling Deep Learning for Materials Discovery, *Nature*, 2023, **624**, 80–87.
7. A. Y.-T. Wang, *et al.*, Machine Learning for Crystal Structure Predictions and Inverse Design, *Chem. Rev.*, 2023, **123**, 12356–12427.
8. J. Schmidt, *et al.*, Recent Advances and Applications of Machine Learning in Solid-State Materials Science, *npj Comput. Mater.*, 2019, **5**, 83.
9. A. Jain and S. P. Ong, *et al.*, The Materials Project: A Materials Genome Approach, *APL Mater.*, 2013, **1**, 011002.
10. K. Choudhary, *et al.*, JARVIS: An Integrated Infrastructure for Data-Driven Materials Design, *npj Comput. Mater.*, 2020, **6**, 173.
11. B. DeCost, *et al.*, A Practical Guide to Microstructure Informatics, *Annu. Rev. Mater. Res.*, 2019, **49**, 1–29.

12. B. P. MacLeod, *et al.*, Self-Driving Laboratory for Accelerated Discovery of Thin-Film Materials, *Sci. Adv.*, 2020, **6**, eaaz8867.
13. F. Häse, L. M. Roch and A. Aspuru-Guzik, Next-Generation Experimentation with Self-Driving Laboratories, *Trends Chem.*, 2019, **1**, 282–291.
14. P. I. Frazier, *A Tutorial on Bayesian Optimisation*, arXiv, 2018, 1807.02811.
15. A. Seko, *et al.*, Prediction of Low-Thermal-Conductivity Materials Using a Machine Learning Method, *Phys. Rev. Lett.*, 2015, **115**, 205901.
16. C. Kim, *et al.*, Active Learning in Materials Science with Benchmarking and Workflows, *npj Comput. Mater.*, 2021, **7**, 110.
17. C. Chen and S. P. A. Ong, Universal Graph Deep Learning Framework for Molecules and Crystals (MEGNet)—Applications and Insights, *Acc. Chem. Res.*, 2022, **55**, 109–119.
18. P. C. St. John, *et al.*, Message-Passing Neural Networks for Polymers, *Macromolecules*, 2019, **52**, 9144–9156.
19. Y. Chen, *et al.*, Inverse Design of Solid Electrolytes with Targeted Ionic Conductivity Using Machine Learning, *Energy Environ. Sci.*, 2021, **14**, 6157–6168.
20. G. N. Simm and M. Reiher, Error Control in Machine-Learning Potentials, *J. Chem. Theory Comput.*, 2021, **17**, 4647–4667.
21. Z. Xiong, *et al.*, Crystal Diffusion Variational Autoencoder for Periodic Material Generation, *Adv. Sci.*, 2023, **10**, 2206230.
22. K. M. Jablonka, *et al.*, Bias, Fairness, and Robustness in Materials Machine Learning, *Chem. Mater.*, 2023, **35**, 4662–4677.
23. M. W. Gaultois, *et al.*, Data-Driven Review of Thermoelectric Materials, *Chem. Mater.*, 2013, **25**, 2911–2920.
24. P. Raccuglia, *et al.*, Machine-Learning-Assisted Materials Discovery from Failed Experiments, *Nature*, 2016, **533**, 73–76.
25. R. E. A. Goodall and A. A. Lee, Predicting Materials Properties without Crystal Structure: Deep Representation Learning from Stoichiometry, *npj Comput. Mater.*, 2020, **6**, 164.

5 Transforming Chemical Manufacturing with AI

5.1 Chapter Overview

Chemical manufacturing—including the synthesis of active pharmaceutical ingredients (APIs) and the production of pharmaceutical products and dosage forms—has always balanced quality, throughput, and cost—now under further constraints of sustainability and regulatory compliance. Artificial intelligence (AI) and robust data infrastructure make it practical to learn directly from production data: to predict quality before it drifts out of specification, to optimise set-points against multiple objectives in real time, and to anticipate failures before they stop the line. Crucially, this is achieved without tearing down validated systems. This chapter shows where AI fits in the plant stack; how to build soft sensors, digital twins and advisory control layers (RTO/MPC/BO); and how to deploy them under GxP expectations (ALCOA+, audit trails, change control). Examples throughout draw on both fine-chemical and pharmaceutical manufacturing, including API crystallisation, excipient handling and dosage-form unit operations such as granulation, tableting, coating and sterile fill–finish.

RSC Foundations No. 7
AI Revolution in Chemistry
By Brian McKew
© Brian McKew 2026
Published by the Royal Society of Chemistry, www.rsc.org

5.2   Core Concepts and Definitions

5.2.1   Where AI Fits in the Plant (Figure 5.1)

A modern plant exposes data through layers:

Field and control layer (PLCs/DCS)—hard real-time control of pumps, valves, heaters, drives.

SCADA/HMI—visualisation and supervisory control; alarms and permissives.

Historians—tag time-series, event frames, calculated/aggregated signals.

MES/ERP—batches, genealogy, scheduling, inventory, recipes (ISA-88/95).[7]

LIMS/ELN/QMS—lab results, release testing, deviations/CAPA; document control.

PAT—inline/online sensors (NIR/IR/Raman/UV–vis), MS, NMR flow probes, particle size/turbidity, calorimetry, conductivity, density, imaging.

AI sits alongside these layers *via* OPC UA/MQTT/REST, consuming raw and contextualised data and returning soft-sensor values, maintenance risk scores or set-point recommendations. In validated, safety-critical loops, AI outputs feed **advisory** RTO or MPC; the DCS retains final authority.[1] This respects 21 CFR Part 11,[9] GAMP 5 and data integrity expectations[8] while still delivering value.

5.2.2   Essential Vocabulary

PAT—inline/online measurements supporting real-time quality assurance.

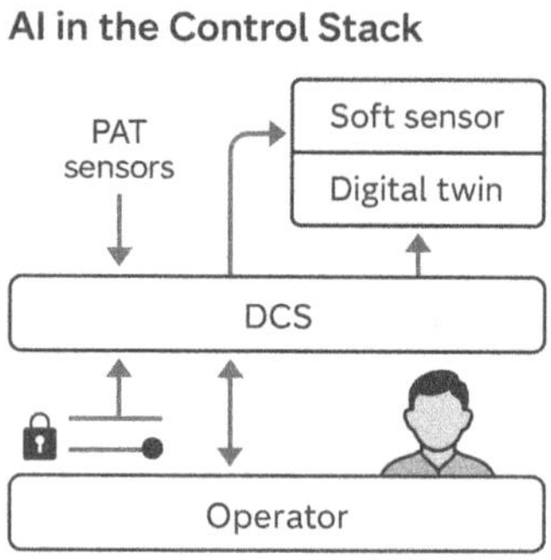

Figure 5.1   AI in the control stack (RTO–MPC–DCS with PAT and soft sensors).

Soft sensor—an ML model inferring a hard-to-measure CQA (*e.g.*, conversion, M̄n, residual solvent) from PAT and process signals.

Digital twin—an empirical or hybrid (first-principles + ML) surrogate for what-if analysis, controller testing and scale-up planning.

MPC—model-predictive control computing future moves under constraints to keep CQAs within bounds.

Run-to-run control—batch-to-batch set-point adaptation.

RTO—real-time optimisation that recommends economically optimal set-points under process and safety constraints.

PdM—predictive maintenance; models forecasting failure risk or remaining useful life for assets.

5.3 Methods and Workflows

5.3.1 Data Readiness and Context Building[2]

1. Contextualise historian data with batch/phase states, equipment modes, and quality milestones (start/charge/heat/hold/cool/unload). Align PAT spectra with process tags *via* a common clock.
2. Clean signals (dropouts, spikes), apply unit normalisation and calibration models; compute engineered features (rolling means, rates, residence-time surrogates).
3. Partition datasets by unit campaign and recipe variant to avoid leakage; maintain time-ordered validation for drift detection.

5.3.2 Soft Sensors for CQAs

Goal—estimate CQAs (*e.g.*, conversion, particle size, residual solvent) continuously.

Inputs—PAT spectra/time-series, actuator settings, temperatures/pressures/flows.

Models—chemometric baselines (PLS/PLS-DA), tree ensembles, 1D CNNs/temporal transformers where justified.

Practice—embed uncertainty (prediction intervals *via* ensembles/conformal), compute applicability domain; schedule recalibration and bias checks after maintenance (*e.g.*, lamp replacement).

Validation—split by campaign or time; confirm on hold-out batches; compare against lab QC with guard-bands.

5.3.3 Digital Twins (Empirical and Hybrid)

Empirical twins—GP/ensemble surrogates mapping set-points to CQAs/energy/throughput.

Hybrid twins—first-principles balances + kinetic models with data-driven corrections for unmodelled phenomena (fouling, non-ideal mixing).

Use—what-if analysis; MPC design and testing; virtual commissioning; scale-up risk assessment (solvent swaps, heat-transfer limits, crystallisation profiles).

5.3.4 Advisory Optimisation and Control (RTO/MPC/BO)[3]

RTO—computes economically optimal set-points subject to safety and quality constraints; pushes recommendations to operators or MPC.

MPC—enforces hard constraints on temperatures/pressures/flows; integrates soft-sensor CQAs as controlled variables; retains operator overrides and permissives.

BO—batch set-point tuning (reaction, crystallisation, separations) with bounded spaces and feasibility/safety classifiers; embeds replication and robustness tests around optima.[4–6]

5.3.5 Compliance and Governance[10]

Data integrity—ALCOA+ across historian, LIMS and model artefacts; immutable audit logs for training data, features, models and deployments.

Change control—record rationale, impacted CQAs, retrain triggers and rollback paths; require sign-off across Quality, Production and Engineering.

Security—role-based access; segregation of duties for model editing *vs.* deployment; network zones for control *vs.* analytics.

5.4 Case Applications

5.4.1 Continuous Pharma (Flow Synthesis)

Soft sensors estimate conversion and impurity; MPC holds CQAs within bounds under feed variability; BO tunes residence time/temperature to reduce E-factor while preserving quality.

5.4.2 Crystallisation with Inline Raman

Twin predicts supersaturation; MPC manipulates temperature and antisolvent addition to steer crystal size distribution; robustness tests probe $\pm\Delta T$ and $\pm10\%$ feed variations.

5.4.3 Polymerisation with Soft Sensors

NIR-based soft sensor estimates $\bar{M}n$ and conversion; RTO/MPC maximise throughput subject to viscosity and quality constraints; scheduled recalibration maintains accuracy.

5.4.4 Separations (Distillation/Filtration)

BO tunes reflux/pressure/temperature or TMP/cycle time to meet purity/throughput/energy targets; feasibility models avoid fouling and flooding regimes.

5.5 Common Pitfalls and How to Mitigate Them

Shortcut variables and leakage—models learn equipment IDs or time-of-day quirks; **fix:** remove non-causal identifiers; validate on future campaigns; perform error forensics.

Unbounded optimisation—recommended set-points violate safety/quality; **fix:** encode hard bounds and interlocks in MPC/RTO; keep BO spaces conservative; require operator confirmation.

Uncalibrated PAT—soft-sensor drift after maintenance; **fix:** calibration standards, drift monitors, and recalibration SOPs; pause/adapt models after lamp/source changes.

Objective mis-specification—optimising proxy (*e.g.*, on-line assay) that fails at QC; **fix:** align objective with released CQAs; include penalties (energy, solvent, off-spec cost).

Governance gaps—models change without records; **fix:** change control, audit trails, versioning; periodic model reviews; defined re-train triggers.

Accessibility and documentation—opaque outputs; **fix:** publish model cards, uncertainty and AD; keep operator job aids and alarm rules simple.

5.6 Implementation Guidance (Step-by-step) (Figures 5.2 and 5.3)

5.6.1 PAT + Soft Sensor Deployment (Batch or Continuous)

Define target CQAs and on-line proxies; perform PAT selection and calibration.

Assemble training data with time alignment and campaign splits.

Fit baseline chemometrics; escalate to ensembles/CNNs only as needed.

Quantify uncertainty and AD; set alarm bands and fallback rules.

Commission in shadow mode; then promote to advisory use; finally integrate with MPC.

5.6.2 Advisory BO for Set-point Tuning (Bounded)

Define safe design space and constraints (*e.g.*, $T \leq 120$ °C; pressure limit; solvent compatibility).

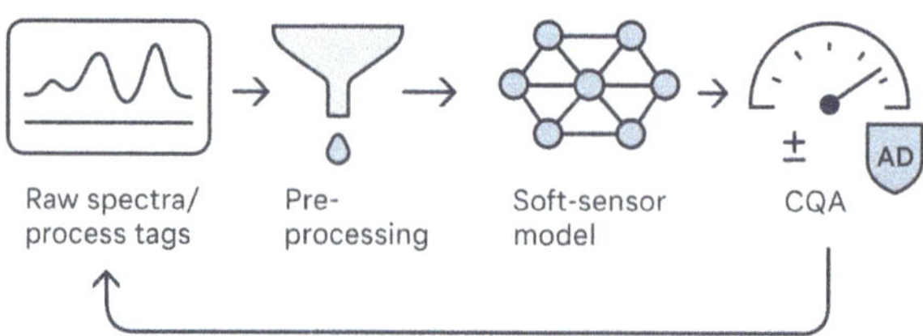

Figure 5.2 From PAT to features to soft-sensor CQAs.

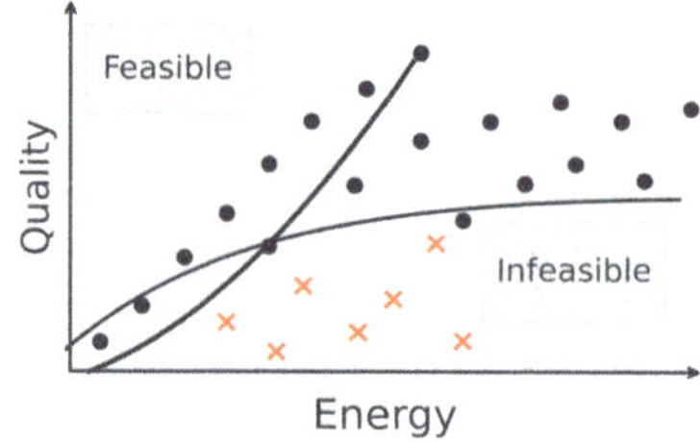

Figure 5.3 Multi-objective BO for energy–quality trade-off.

Seed with space-filling or historical data; include replication (10–20%) to quantify noise/drift.

Run closed-loop batches; enforce diversity; add robustness tests near optima.

Capture improvements in yield/energy/quality; write back to recipes with change control.

5.6.3 MPC with Soft-sensor CQAs

Identify controlled variables (lab CQAs proxied by soft sensors), manipulated variables and constraints.

Design and test controllers against the twin; validate in plant with staged authority.

Enable advisory RTO to update targets; maintain operator overrides and permissives.

5.6.4 Scale-up and Solvent Swap Risk Mitigation

Use a hybrid twin to evaluate heat/mass transfer and impurity risks under candidate solvents/routes.

Plan experiments *via* BO with feasibility models; confirm under worst-case feeds and utilities.

5.7 Key Takeaways

AI augments validated plants without displacing the DCS: start advisory, bound authority, and keep operators in the loop.

Soft sensors and twins unlock continuous visibility of CQAs and robust what-if analysis; combine with MPC/RTO for constraint-aware control.

BO provides safe, sample-efficient set-point tuning for batch and continuous units; always encodes feasibility and replication.

Compliance is a design constraint: ALCOA+, audit trails, model cards, and change control are non-negotiable.

Sustainability can be a first-class objective alongside quality and throughput: include energy, solvent and emissions in optimisation.

References

1. D. E. Fitzpatrick and S. V. Ley, Engineering Chemistry: Integration of Flow, Analytics and Control, *Org. Process Res. Dev.*, 2021, **25**, 1864–1884.

2. W. H. Green, *et al.*, Data and Uncertainty in Chemical Kinetics and Catalysis, *ACS Catal.*, 2021, **11**, 10672–10696.
3. F. Häse, L. M. Roch and A. Aspuru-Guzik, Next-Generation Experimentation with Self-Driving Laboratories, *Trends Chem.*, 2019, **1**, 282–291.
4. B. P. MacLeod, *et al.*, Self-Driving Laboratory for Accelerated Discovery of Thin-Film Materials, *Sci. Adv.*, 2020, **6**, eaaz8867.
5. B. J. Shields, J. Stevens and L. Cronin, Bayesian Reaction Optimisation as a Tool for Chemical Synthesis, *Nat. Rev. Chem.*, 2021, **5**, 315–330.
6. G. Tom, *et al.*, Self-Driving Laboratories for Chemistry and Materials Science, *Chem. Rev.*, 2024, **124**, 5905–5977.
7. ISA-88; ISA-95, Instrumentation, Systems, and Automation Society Standards for Batch Control and Enterprise-Control Integration.
8. GAMP 5: A Risk-Based Approach to Compliant GxP Computerised Systems, ISPE, 2nd edn, 2022.
9. U.S. FDA. 21 CFR Part 11—Electronic Records; Electronic Signatures.
10. ICH Q8(R2), Q9(R1), Q10, Q13, Q14—Pharmaceutical Development, Quality Risk Management, Pharmaceutical Quality System, Continuous Manufacturing, Analytical Procedure Development.

6 Enhancing Chemical Safety and Toxicology Through AI

6.1 Chapter Overview

Chemical safety and toxicology have always been about anticipation: foreseeing harmful effects before people, communities, or ecosystems bear them. Historically, that anticipation relied on animal tests, targeted *in vitro* assays, expert judgement, and conservative defaults. These approaches remain important, but they can be slow, costly, and ethically constrained, and they struggle to keep pace with the sheer number of substances in commerce and the mixture and life-cycle realities of modern chemistries. Over the past decade—and decisively in recent years—artificial intelligence (AI) has become a practical way to extend and focus safety science: to predict toxicity earlier, to integrate diverse lines of evidence coherently, to reduce reliance on animals, and to make environmental monitoring more timely and informative.

This chapter provides a chemistry-first orientation to AI in safety and toxicology. We clarify foundational concepts—QSAR/QSPR, read-across, applicability domain (AD), and new approach methodologies (NAMs)—and show how AI enriches each. We then present workflows that practising chemists and safety scientists can adopt: from defendable QSARs for critical endpoints (*e.g.*, skin sensitisation, hERG, Ames, DILI), to multi-task property models, to mechanistically anchored ensembles that combine *in vitro* bioactivity, structural alerts,

RSC Foundations No. 7
AI Revolution in Chemistry
By Brian McKew
© Brian McKew 2026
Published by the Royal Society of Chemistry, www.rsc.org

omics signatures, adverse outcome pathways (AOPs), and physiologically based pharmacokinetic (PBPK) models. We cover uncertainty quantification (UQ) and calibration, domain checks, and how to structure weight-of-evidence (WoE) arguments that satisfy regulators and internal governance. Case studies illustrate where AI has already helped reduce animal use, triage regulatory workloads, and sharpen surveillance of environmental hazards.

Finally, we offer implementation guidance—what data to assemble, how to split and validate, how to document model cards and change control, and how to embed ethics, fairness, and transparency in day-to-day decisions. Throughout, we emphasise that AI does not replace toxicology; it amplifies it—by making better use of the information we already have, by pointing efficiently to the next most informative experiment, and by knitting disparate signals into coherent risk narratives.

6.2 Core Concepts and Definitions

6.2.1 From QSAR to AI-enabled Safety Science (Figure 6.1)

QSARs/QSPRs (Quantitative Structure–Activity/Property Relationships) use chemical structure to predict biological or environmental

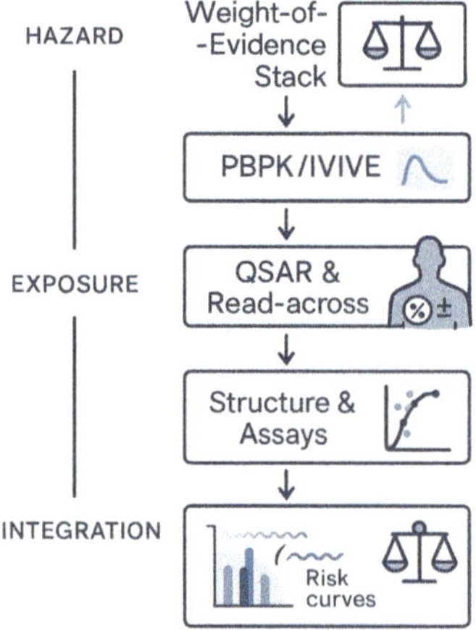

Figure 6.1 AI-enabled risk-assessment stack.

outcomes (*e.g.*, mutagenicity, skin sensitisation, bioaccumulation). AI extends QSAR with non-linear learners (gradient-boosted trees, deep neural networks, graph neural networks—GNNs) that capture complex structure–activity relationships beyond hand-crafted descriptors.

Read-across infers properties of a target chemical from an analogue set. AI improves read-across by (i) learning more informative chemical embeddings, (ii) quantifying uncertainty around the inference, and (iii) combining structure with bioactivity fingerprints (from high-throughput screening) so similarity is defined in a mechanistically relevant space.

Applicability domain (AD) defines where a model's predictions are reliable. Modern practice uses distance in learned latent space, leverage in descriptor space, conformal prediction sets, and model ensembles to flag out-of-domain inputs and attach calibrated confidence to each prediction.

New approach methodologies (NAMs)[15] include *in vitro* assays, computational models, *in silico* exposure tools, and organ-on-chip/ microphysiological systems intended to complement or replace animal tests. AI helps (i) integrate NAM outputs, (ii) account for assay drift and protocol variance, and (iii) map bioactivity to AOP frameworks that connect molecular initiating events to apical outcomes.

Weight-of-evidence (WoE) is the disciplined process of assembling, scoring, and integrating heterogeneous evidence streams (QSARs, read-across, assays, exposure models, analogues, literature) into a transparent argument for a conclusion. AI supports evidence synthesis (*e.g.*, probabilistic graphical models) and NLP for literature extraction, within governance that keeps the human toxicologist accountable.

6.2.2 Exposure and Risk, Not Just Hazard

Predicting hazard (*e.g.*, "mutagenic: yes/no") is insufficient; risk depends on exposure and toxicokinetics. AI assists by:

- learning exposure predictors[23] from product-use patterns, emissions inventories, and environmental fate models;
- linking *in vitro* potency to *in vivo* equivalent doses *via* PBPK models whose parameters are estimated or bounded with data-driven surrogates (IVIVE/QIVIVE);
- estimating mixture effects *via* multi-task models that share representations across chemicals and endpoints while propagating uncertainty.

6.2.3 Environmental Monitoring and Impact

Safety extends beyond human health. AI supports environmental toxicology by

- interpreting remote-sensing and sensor-network time-series (*e.g.*, harmful algal blooms, turbidity plumes, oil slicks);
- detecting and classifying microplastics *via* image analysis and spectral fingerprints (FTIR/Raman);
- estimating bioaccumulation and food-web transfer using graph-based ecological models and spatial statistics.

6.3 Methods and Workflows

6.3.1 Defendable QSAR Development for Critical Endpoints (Figure 6.2)

(a) Decide the endpoint and audience.

   Specify the decision (screening, prioritisation, waiving, or regulatory submission) and align to OECD principles:[1] defined endpoint, unambiguous algorithm, defined AD, appropriate goodness-of-fit/robustness/predictivity, and mechanistic interpretation where feasible.

(b) Assemble and clean the dataset.

   Curate structures (tautomer/salt/solvent handling; stereochemistry; charge state at pH 7.4 where relevant).[6]

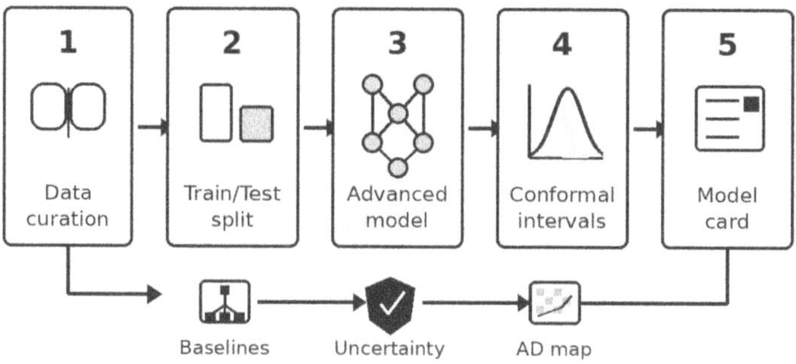

Figure 6.2 Defendable QSAR workflow with AD and calibration.

De-duplicate across sources; harmonise assay conditions (cells/strains/media/time/solvent/controls) and units.

Resolve conflicts (same compound, different labels) *via* majority vote, weighted by assay reliability, or exclude.

Retain uncertainty for continuous endpoints (*e.g.*, replicate variance for log K_ow, IC_{50}).

(c) Choose representations and baselines.

Start with fingerprints (ECFP/FCFP), physicochemical descriptors (TPSA, HBD/HBA, c Log P, pK_a), and structural alerts.[8]

Add graph representations (GNNs) if data volume and endpoint complexity warrant; record descriptor parameters and provenance.

(d) Split to match deployment.

For medicinal-chemistry-like SAR, use scaffold or time-based splits to avoid analogue leakage.

For diverse regulatory QSARs, use cluster-based or stratified chemotype splits.

Keep an external test set untouched until the final evaluation.

(e) Model families and calibration.

Compare regularised linear/logistic, random forests/gradient boosting, and GNNs/transformers.

Report ROC-AUC/PR-AUC (classification) and RMSE/MAE (regression) with calibration (Brier score, reliability curves).

Use conformal prediction[5] for per-compound intervals/sets; validate coverage on validation folds.[22]

(f) AD and interpretability.

Define AD *via* descriptor/latent-space distance; set reject rules for OOD.

Provide mechanistic context: align attention/attribution with alerts/toxicophores; relate features to AOP key events.

(g) Documentation and governance.

Produce a model card: purpose, endpoint, data sources, splits, features, algorithms, metrics, AD, UQ, limitations, retrain triggers.

Archive training artefacts (seeds, library versions, descriptor settings) for reproducibility.

6.3.2 Read-across with AI Assistance[3]

(a) Define the analogue space.[19] Combine structural similarity with bioactivity (docking/HTS fingerprints) and metabolic similarity (predicted biotransforms). Note toxicophores and reactive moieties.

(b) Select analogues quantitatively. Use k-NN in a joint embedding; weight neighbours by domain relevance (assay, organism, exposure route).

(c) Quantify uncertainty. Bootstrap analogue sets to obtain confidence intervals; triangulate with QSAR predictions and *in vitro* surrogates in a WoE framework.

(d) Mechanistic annotation. Map analogues to AOP key events; flag metabolic activation risks (*e.g.*, quinone imine formation). Explain why each analogue is informative beyond a raw similarity score.

6.3.3 Multi-task and Federated Models

(a) Multi-task learning. Train a shared representation for multiple toxicological endpoints (*e.g.*, hERG, CYP inhibition, Ames, skin sensitisation, DILI).[7] Benefits: data efficiency and improved calibration on sparse endpoints. Guard against negative transfer with task weighting or per-task heads.

(b) Federated learning. Where data cannot be centralised, train across partners *via* federated averaging with privacy safeguards (secure aggregation, access control). Verify that the federated model remains calibrated on each site's hold-out; document data-sharing governance.

6.3.4 Mechanistically Anchored Ensembles

(a) *In vitro* bioactivity panels. Use Tox21/ToxCast-type profiles[9] as feature vectors; learn patterns that connect pathway activation (*e.g.*, ER agonism) to endpoints within AOP frameworks.

(b) Omics signatures. Derive gene-set enrichments from RNA-seq/proteomics; build biologically informed embeddings; link to AOP key events (KEs)[20] to improve interpretability.

(c) PBPK + IVIVE.[13] Estimate PBPK parameters[14] (partition coefficients, clearance) with surrogates; run *in vitro–in vivo* extrapolation (IVIVE/QIVIVE) to obtain human-equivalent doses from *in vitro* potencies, with uncertainty bands.

(d) Weight-of-evidence integration. Combine QSAR, read-across, *in vitro* data, and exposure models using a Bayesian evidence graph[4] or similar framework; report a posterior probability of concern and trace contributions.

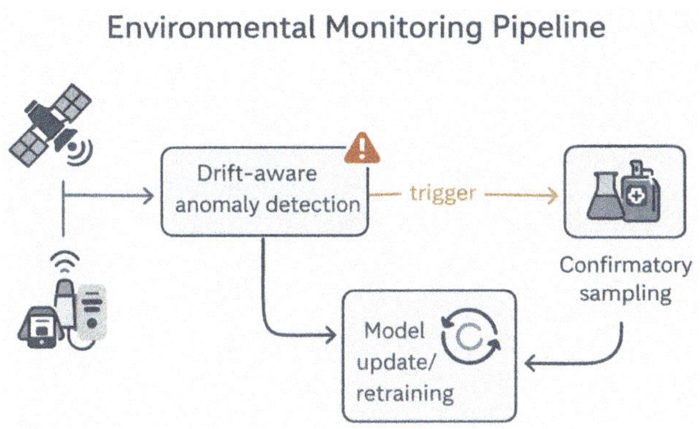

Figure 6.3 Environmental monitoring pipeline.

6.3.5 Environmental Fate and Effects (Figure 6.3)

(a) Fate and transport. Train surrogates for sorption, degradation (hydrolysis, photolysis, biodegradation), and volatilisation; feed outputs (with UQ) into multimedia fate models to estimate compartment concentrations.

(b) Bioaccumulation and trophic transfer.[24] Predict BCF/BMF using structure and context; apply graph-based ecological networks for trophic propagation and concentration estimates with uncertainty.

6.3.6 Monitoring

Remote sensing: detect algal blooms and plumes using spectral indices and anomaly detection.

Sensor networks: apply drift-aware ML to long time-series from low-cost sensors; compensate for temperature/humidity; flag events for confirmatory sampling and lab analysis.[11]

6.4 Case Studies and Recent Signals (2024–2025)

6.4.1 Case 1—Deep Learning Lifts Consensus[2] on Tox21 Endpoints

In Tox21-style panels,[10] deep neural networks exceeded many classical learners on pathway-level endpoints. Multi-task models further improved calibration for sparser assays while preserving performance on dense tasks. Recent refinements (2024) add conformal prediction, enabling defensible triage rules (*e.g.*, accept high-confidence negatives, route ambiguous compounds to assay).

Pattern: Multi-task DNN/GNN + conformal UQ + AD checks + assay-aware splits → defendable NAM-first screening.

6.4.2 Case 2—Skin Sensitisation *via* NAMs:[18] DPRA + KeratinoSens + QSAR

For skin sensitisation, defined approaches combine[12] DPRA peptide reactivity, KeratinoSens/ARE-Nrf2 activation, and QSAR/read-across. AI calibrates decision boundaries, resolves conflicts with a Bayesian integrator, and prioritises confirmatory work. Regulators have accepted such submissions in defined contexts, reducing LLNA reliance.

Pattern: pre-specified rules + transparent integrator + UQ bands → credible dossiers.

6.4.3 Case 3—DILI Composite Models[21] for Risk Triage

Drug-induced liver injury (DILI) remains challenging. Composite models merge structure-based learners, *in vitro* stress panels (mitochondrial liability; BSEP inhibition), and clinical covariates, while rejecting OOD scaffolds. Idiosyncratic DILI is not perfectly predictable, but composites raise negative predictive value, focusing *in vivo* studies where needed.

Pattern: composite endpoints + multi-source integration + OOD rejection to reduce false reassurance.

6.4.4 Case 4—Microplastic Identification and Load Estimation

FTIR/Raman spectral imaging + CNNs accelerates microplastic detection. Active learning cuts labelling burden and adapts to new polymers/weathering states. Pipelines integrate segmentation, polymer class prediction, and size/mass estimation with quality flags for ambiguous spectra.

Pattern: standardised pre-processing + AL-assisted curation + calibrated outputs → scalable monitoring.

6.5 Common Pitfalls and How to Mitigate Them

6.5.1 Data Leakage and Optimistic Validation

Risk: Random splits allow near-duplicates; lab/vendor artefacts leak into features.

Mitigation: Use scaffold/time/site-wise splits; scrub operational identifiers; preserve an external test until final evaluation.

6.5.2 Out-of-domain (OOD) Deployment

Risk: Applying a drug-like model to polymers, PFAS-like chemistries, or metal complexes.

Mitigation: Define AD (descriptor/latent distances); implement reject rules; couple predictions with UQ; collect targeted data to expand domain.

6.5.3 Conflating Hazard with Risk

Risk: Treating *in vitro* activity as *in vivo* hazard; ignoring exposure/ kinetics.

Mitigation: Pair hazard with PBPK/IVIVE; report human-equivalent doses with bands; model realistic exposure scenarios.

6.5.4 Single-surrogate Overreach

Risk: Relying on docking or a single HTS assay or an uncalibrated QSAR for high-stakes calls.

Mitigation: Use ensembles and WoE; require orthogonal evidence before escalation.

6.5.5 Uninterpretable Models in Critical Contexts

Risk: Strong metrics but no mechanistic defence.

Mitigation: Provide attributions/attention aligned with alerts/AOPs; include counterfactuals; retain simple challenger models.

6.5.6 Neglecting Uncertainty

Risk: Point estimates presented as certainties.

Mitigation: Deploy conformal/ensemble UQ; publish calibration plots; set thresholds by error-cost.

6.5.7   Governance Gaps

Risk: No model cards; ad-hoc retraining; no change control.

Mitigation: Implement model lifecycle management: cards, versioning, ALCOA+ trails, periodic performance reviews.

6.5.8   Ethics and Fairness

Risk: Training-set bias (under-representation of industrial chemistries or non-OECD assays) drives inequitable decisions.

Mitigation: Audit coverage; diversify data sources; report coverage distributions; include fairness checks.

6.6   Implementation Guidance (Step-by-step)

6.6.1   Build a Defendable QSAR (Regulatory-relevant)

Frame the decision: Screening, waiving, or prioritisation; define false-negative/positive tolerance.

Curate data: Consolidate public/internal sets; harmonise protocols; normalise structures.[17]

Split deliberately: Scaffold/time/site-wise; hold out an external test.

Train baselines first: Regularised linear/logistic + trees with interpretable features/alerts.

Escalate cautiously: Add GNNs/transformers only if data and endpoint justify.

Quantify uncertainty: Conformal/ensembles; verify coverage on validation folds.

Define AD/reject: Declare validity region; specify handling outside AD.

Document: Write a model card; record seeds/libs/descriptor parameters; pre-register retrain triggers (*e.g.*, +20% diverse data).

6.6.2   Perform Read-across with AI Support

Assemble analogues by structure, bioactivity, and metabolic similarity.

Score relevance transparently; retain provenance for each analogue.

Quantify uncertainty *via* bootstrap sets; triangulate with QSAR and *in vitro* surrogates.

Explain with mechanistic anchors (alerts/AOPs) why each analogue is probative.

6.6.3 Integrate Hazard and Exposure (Risk-centred)

IVIVE/QIVIVE: Link *in vitro* potency to dose *via* PBPK surrogates (partitioning, clearance); propagate uncertainty.[25]

Risk curves: Combine hazard probabilities with exposure distributions; report probability of exceeding effect thresholds.

Scenario analysis: Consumer, worker, environmental; present margin of safety with uncertainty bands.

6.6.4 Environmental Impact Workflow[16]

Predict fate parameters (sorption, persistence, volatilisation) and bioaccumulation with UQ; feed into the multimedia fate model.

Deploy monitoring: Remote sensing for blooms/plumes; sensor networks with drift-aware ML.

Close the loop: Confirm anomalies with targeted sampling; retrain thresholds quarterly.

6.6.5 Documentation and Governance

Model cards: Purpose; data sources; splits; algorithms; metrics; UQ/AD; limitations; retrain schedule.

Change control: Version datasets/models; require approvals for updates; keep an audit trail.

Transparency pack: Calibration/reliability plots; AD maps; example counterfactuals; WoE narrative template.

Ethics/fairness: Record coverage across chemotypes/use-classes; document fairness checks; make model limits and governance transparent.

6.7 Key Takeaways

Start with the decision: Define endpoint, audience, and error tolerance; design QSAR/read-across accordingly.

Validate like deployment: Scaffold/time/site-wise splits, conformal UQ, and AD; keep an external test set.

Integrate hazard and exposure: Pair *in vitro* hazard with PBPK/IVIVE and exposure; present risk curves with bands.

Prefer ensembles and WoE: Combine QSAR/read-across, bioactivity, and AOP anchors; avoid single-surrogate overreach.

Reduce animal use with NAMs: Calibrate and integrate DPRA/KeratinoSens/HTS; pre-specify decision criteria.

Monitor ethically and fairly: Audit coverage; document fairness checks; make model limits and governance transparent.

Environment is part of safety: Apply AI to fate/effects and monitoring; close the loop with confirmatory sampling and retraining.

Govern lifecycle: Model cards, change control, ALCOA+ trails, retrain triggers → reliable, audit-ready safety AI.

References

1. OECD, Principles for the Validation, for Regulatory Purposes, of (Q)SAR Models, OECD, Paris, 2004.
2. A. Mayr, *et al.*, DeepTox: Toxicity Prediction Using Deep Learning, *Front. Environ. Sci.*, 2016, **3**, 80.
3. C. Helma, *et al.*, Read-across with REACH: Consequences for Scientific Standards, *Regul. Toxicol. Pharmacol.*, 2018, **98**, 162–169.
4. J. Jaworska, *et al.*, Bayesian Data Analysis for Weight-of-Evidence Integration in Regulatory Toxicology, *Regul. Toxicol. Pharmacol.*, 2010, **58**, 220–226.
5. U. Norinder and S. Boyer, Conformal Prediction for Reliable Toxicity Classification, *J. Chem. Inf. Model.*, 2017, **57**, 459–468.
6. L. C. Blum and J.-L. Reymond, 970 Million Drug-like Small Molecules for Virtual Screening, *J. Am. Chem. Soc.*, 2009, **131**, 8732–8733.
7. H. Li, C. W. Yap and C.-Y. Ung, *et al.*, Predicting hERG Potassium Channel Inhibition with Machine Learning, *Bioinformatics*, 2021, **37**, 4710–4718.
8. N. Greene, *et al.*, Structural Alerts for Toxicity: Mechanisms, Toxicophores and Warheads, *Drug Discovery Today*, 2020, **25**, 2259–2277.
9. D. J. Dix, *et al.*, The ToxCast Programme for Prioritising Toxicity Testing of Environmental Chemicals, *Toxicol. Sci*, 2007, **95**, 5–12.
10. R. Kavlock, *et al.*, Tox21: A Vision and a Strategy, *Toxicol. Sci*, 2009, **107**, 307–313.
11. B. I. Escher, *et al.*, From Chemical Mixtures to Effect-Based Water Quality Monitoring, *Environ. Sci. Technol.*, 2018, **52**, 9767–9775.
12. OECD, Guidance on Reporting of Defined Approaches and IATA; Series No. 255 (2016; 2021 update).
13. B. A. Wetmore, Quantitative In Vitro–to–In Vivo Extrapolation: A Critical Part of Next-Generation Risk Assessment, *Toxicol. Sci*, 2015, **148**, 325–336.
14. A. Paini, *et al.*, Next-Generation PBK Models in Chemical Risk Assessment, *Toxicol. In Vitro*, 2019, **59**, 1–3.
15. V. Kandarpa, *et al.*, Microphysiological Systems for Toxicity Testing: A Roadmap, *ALTEX*, 2022, **39**, 3–21.
16. European Commission, Chemicals Strategy for Sustainability: Towards a Toxic-Free Environment, 2020.
17. U.S. EPA, CompTox Chemicals Dashboard (Accessed 2024). https://comptox.epa.gov/dashboard.
18. Council of Europe/EDQM, Skin Sensitisation: Non-Animal Defined Approaches, 2020–2024 guidance updates.
19. T. Luechtefeld, C. Rowlands and T. Hartung, Big-Data Read-Across for Acute Systemic and Topical Toxicity, *Altex*, 2018, **35**, 139–150.
20. OECD, Adverse Outcome Pathways (AOP) Knowledge Base (Accessed 2024). https://aopkb.oecd.org.
21. M. Chen, *et al.*, DILIrank: A Reference Drug List Ranked by DILI Risk, *Drug Discovery Today*, 2016, **21**, 648–653.

22. U. Norinder, S. Boyer and M. Eklund, Conformal Prediction in QSAR—AD with Error Control, *J. Chemom.*, 2015, **29**, 233–244.
23. J. F. Wambaugh, *et al.*, High-Throughput Heuristics for Chemical Exposure, *Environ. Sci. Technol.*, 2014, **48**, 12760–12767.
24. J. A. Arnot and F. A. P. C. Gobas, Food-Web Bioaccumulation Model for Organic Chemicals, *Environ. Toxicol. Chem.*, 2004, **23**, 2343–2355.
25. OECD, Guidance on the Characterisation, Validation and Reporting of PBK Models, 2021.

7 Navigating Ethical Considerations and Regulations

7.1 Chapter Overview

Artificial intelligence (AI) is accelerating the way chemists design, make, measure, and decide. Yet the same properties that make AI powerful—its ability to learn from data, to generalise, and to automate—create new obligations in safety-critical and regulated settings. In chemistry, models influence choices with health, environmental, and economic consequences: which molecule to make, which waste stream to treat, which process to release, which community to monitor. If those models are biased, opaque, fragile, or poorly governed, they can amplify harm rather than reduce it.

This chapter provides a practical, chemistry-first guide to ethical AI and regulation. We clarify the core concepts—bias, fairness, transparency, accountability, and human oversight—and translate them for laboratory research, manufacturing, safety/toxicology, and environmental monitoring. We then walk through workflows to manage model risk end-to-end: governance-by-design, data and model cards, applicability domain (AD) and uncertainty quantification (UQ), HAZOP-style hazard analysis for AI-enabled operations, change control, incident response, and supplier risk for third-party models.

RSC Foundations No. 7
AI Revolution in Chemistry
By Brian McKew
© Brian McKew 2026
Published by the Royal Society of Chemistry, www.rsc.org

We situate these practices within the current regulatory landscape (*e.g.*, the EU AI Act and sectoral requirements such as GxP/ICH in pharma, REACH/CLP for chemicals, 21 CFR Part 11/GAMP 5 for electronic records, NIST/OECD/ISO guidance for trustworthy AI), explaining what they mean for day-to-day choices in chemistry (Figures 7.1–7.3).

The emphasis throughout is on responsibility that scales: simple checklists and artefacts that slot into existing quality management systems (QMS), ALCOA+ data integrity, and ISA-88/ISA-95 process governance. Implemented early, these steps keep AI productive and auditable—so teams can move quickly without compromising safety, fairness, or compliance.

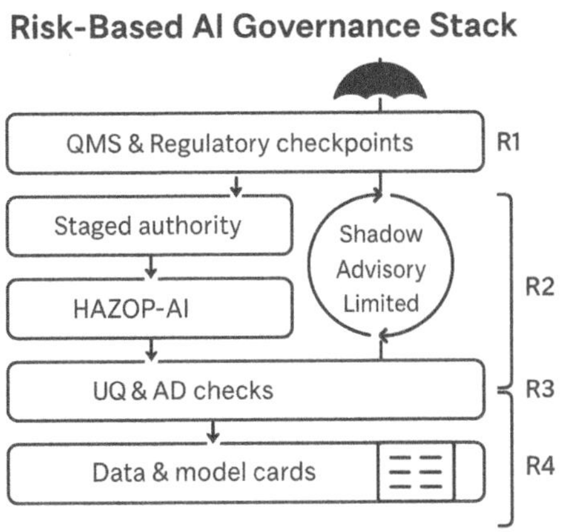

Risk-Based AI Governance Stack

Figure 7.1 Risk-based AI governance stack for chemistry.

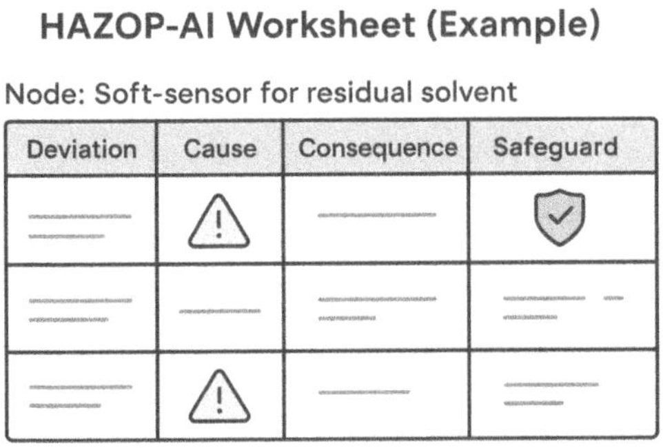

HAZOP-AI Worksheet (Example)

Node: Soft-sensor for residual solvent

Deviation	Cause	Consequence	Safeguard

Figure 7.2 HAZOP-AI worksheet (example).

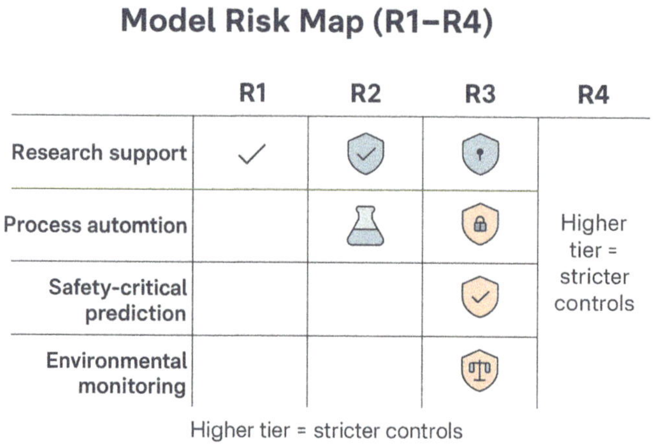

Model Risk Map (R1–R4)

	R1	R2	R3	R4
Research support	✓	🛡	🛡	
Process automtion		⚗	🛡	Higher tier = stricter controls
Safety-critical prediction			🛡	
Environmental monitoring			🛡	

Higher tier = stricter controls

Figure 7.3 Model risk map (R1–R4) for chemistry use-cases.

7.2 Core Concepts[13] and Definitions

7.2.1 Bias, Fairness, and Representativeness (Chemistry-specific)

Sampling bias: Training sets over-represent drug-like scaffolds, specific assay conditions, or Western sites/instruments; models then underperform on industrial chemistries, novel solvents, or low-resource contexts.

Measurement bias: Systematic differences in assay protocols, instrument calibration, or preprocessing (*e.g.*, spectral baselines) leak artefacts into features; models "learn the instrument" rather than the chemistry.

Label bias: Endpoints stitched from heterogeneous sources (*e.g.*, mixed DFT levels, lab protocols, historical heuristics) embed systematic error that a model cannot unlearn.

Fairness in chemistry: It means consistent performance across chemotypes, process windows, sites/operators, communities in environmental monitoring, and demographic groups in human sampling (*e.g.*, biomonitoring)—with documented limits and mitigations.

Practical tests: Report metrics by stratum (chemotype clusters, site/ instrument, substrate class, matrix) and display coverage maps for training *vs* intended use. Attach confidence *via* UQ/AD; trigger additional testing where gaps are large.

7.2.2 Transparency and Explainability[16]

Transparency concerns documentation: Who built what, with which data, under which assumptions; what changed; who approved it.

Explainability is reason-giving: Why did the model rank these candidates or adjust those set-points? In chemistry, explanations must be physically plausible and actionable (*e.g.*, atom-level attributions aligned with toxicophores; reaction features mapping to rate-controlling steps; spectral bands tied to chemical moieties; process drivers consistent with first-principles narratives).[18]

Practical rule: Pair every high-stakes model with (i) a model card, (ii) counterfactuals ("what small change flips the decision?"), and (iii) challenger baselines that are simpler but interpretable.

7.2.3 Accountability and Human Oversight

Human-in-the-loop: People authorise actions, set operating envelopes, and decide when to override the model.

Lines of responsibility: The model owner (technical) and the process owner (operational) co-sign deployments; a QMS change-control board approves updates; incident response defines who stops what and when.

Auditability: Every recommendation and set-point change is attributable (ALCOA+) with input snapshot, model version, seed/ configuration, and confidence attached.

7.2.4 Safety Envelopes for Autonomy

Automation in chemistry is never "hands-off"; it is risk-bounded. Every AI-driven loop must declare: (i) hard constraints (temperature/ pressure/UV/flow limits; exclusion lists for reagents/solvents), (ii) trip conditions (interlocks/watchdogs), (iii) safe-halt/fallback states, and (iv) authority levels (shadow $\rightarrow$ advisory $\rightarrow$ limited closed loop).

7.2.5 Model Risk and Categorisation

Not all models are equal. A risk taxonomy supports proportionate controls:

R1 (informational): Literature assistants, search, summarisation (low operational risk; privacy/IP concerns).

R2 (design triage): Property/QSAR predictors, virtual screening, generative ideation (medium risk; synthesis/process/safety guardrails).

R3 (quality-critical): Soft sensors for CQAs, environmental anomaly detectors (high risk; UQ/AD, calibration, monitoring).

R4 (safety-critical/control): RTO/MPC set-point advisors, autonomous lab loops (very high risk; safety envelopes, HAZOP, staged authority).

Controls scale with category (see governance-by-design).

7.3 Regulatory Landscape (High-level)[5]

EU AI Act[1] (2024): A risk-based horizontal law prohibited, high-risk, limited-risk, and minimal-risk categories; transparency duties for general-purpose AI (GPAI) and foundation models; phased application 2025–2027. Chemistry use-cases that influence health/safety (*e.g.*, release testing, plant control) may be high-risk, triggering documentation, monitoring, and oversight obligations.

ISO/IEC: 23894:2023[2] (AI risk management) and 42001:2024[3] (AI management systems) provide management-system scaffolding; IEC 62443 covers industrial cybersecurity.

NIST AI RMF 1.0 (2023):[4] A voluntary U.S. framework to govern, map, measure, and manage AI risks; it dovetails with quality systems.

OECD AI Principles (2019): Human-centred values, transparency, robustness, accountability—widely referenced.

Sectoral:[6,8] Pharmaceutical GxP/ICH (Q8/Q9/Q10/Q14),[9–11] 21 CFR Part 11,[7] GAMP 5 (2nd ed.); chemicals REACH/CLP/TSCA;[20,21] environmental permits and monitoring rules.[22] These do not target AI specifically but dictate data integrity, validation, and change control.

7.4 Methods and Workflows

7.4.1 Governance-by-design (Lifecycle Template)

Treat every model like an instrument under QMS control. The following artefacts and gates integrate with most chemistry organisations:

Model proposal (Gate 0): Decision to support (screening, release, set-point advice); risk category (R1–R4); intended use/users and off-label exclusions; ethical/stakeholder context.

Data and design (Gate 1): Data cards (sources, protocols, coverage, known biases, legal basis, retention, privacy/IP); pre-registered splits (scaffold/time/site) and metrics; HAZOP-AI initiation; access control/ redaction.

Build and validate (Gate 2): Baselines (linear/trees/chemometrics) first; UQ/AD calibrated (Brier/reliability); stratified performance (chemotypes/sites/matrices); fairness checks; interpretability (attributions/counterfactuals/mechanistic alignment); external test locked.

Pre-deployment (Gate 3): Model card; SOPs for operation/monitoring/escalation; cyber review (interfaces, edge/cloud, encryption); human factors (HMI design, warnings, confidence display).

Staged rollout (Gate 4): Shadow → advisory → limited closed loop under safety envelope; parallel SPC/MSPC for cross-checks; operator sign-off logging.

Operate and monitor (Gate 5): KPIs (MAE/ROC, coverage, reject rate, alarm load, adoption); drift monitors; retrain thresholds; incident playbooks; near-miss logs.

Change control (Gate 6): Impact assessment, approvals, version bump, rollback plan; defined re-validation scope.

Retire/replace (Gate 7): Archive models/data/decisions; risk review; lessons learned.

7.4.2 HAZOP-AI: Hazard Analysis for AI-enabled Chemistry

Blend process safety and model risk: For each node (*e.g.*, soft sensor for residual solvent, BO advisor for distillation energy, autonomous crystalliser):

Deviations: Wrong prediction, over-confident OOD, delayed inference, conflicting sensors, drift, cyber manipulation, LLM hallucination in SOP drafting.

Causes: Instrument drift, feedstock change, unmodelled regime, data pipeline break, poisoned input, version mismatch.

Consequences: Off-spec product, unsafe conditions, data breach, environmental exceedance.

Safeguards: UQ/AD rejects, hard constraints, interlocks, MPC final control, two-person approval for escalations, HMI confidence display.

Actions: Safe-halt, re-calibration, retraining, root-cause analysis, security patching.

Evidence: Model card references; validation reports; cybersecurity asset list.

7.4.3 Data Governance and Documentation

Data cards[15] capture: Provenance, consent/legal basis (if human data), coverage maps, quality (missingness/outliers), preprocessing, known biases, retention/disposal. Model cards[14] record: design intent, training data references, splits, metrics by stratum, calibration plots, AD definition, UQ method, limits, and retrain triggers.

Good record keeping (ALCOA+): Store raw spectra/time-series and configs; log training seeds, library versions, and hashes; attach attribution for every operational decision (who, when, inputs, confidence).

7.4.4 Uncertainty and Applicability Domain as First-class Controls

Conformal prediction (classification/regression) yields per-sample sets/intervals with guaranteed error rates (under exchangeability). Display intervals in the HMI; route ambiguous cases to human review or confirmatory assays.

Ensembles/evidential models provide complementary epistemic/aleatoric signals; track reject ratios and their trend.

AD *via* distance in latent and descriptor spaces; maintain OOD dashboards by chemotype/site/process window.

7.4.5 Security and Privacy in Operational Technology (OT)

Boundary control: Segment OT/IT;[12] strict allow-lists; MFA for remote access.

Edge inference where latency/safety matter; zero-trust for cloud calls; fail-closed on timeouts.

Data minimisation: avoid personal data in logs; mask sensitive supplier data; scrub secrets from LLM prompts.

7.4.6 Supplier and Foundation-model Risk

Third-party models (LLMs, diffusion, protein/materials FMs): Require supplier attestations (training data sources, licences, safety filters), domain benchmarks, and guardrails (prompt templates, retrieval-augmented generation with curated corpora, no direct execution of generated code/procedures).

Content provenance: Use watermarks/signatures where available; keep attribution when generated artefacts enter the QMS.

7.5 Case Studies and Recent Signals (2024–2025)

7.5.1 Signal 1—EU AI Act Programmes Stand Up AI Management Systems

European labs and plants initiated AI management systems aligned with ISO/IEC 42001:2024[3] and NIST AI RMF to prepare for the Act's risk-based obligations. High-risk chemistry use-cases (*e.g.*, release-affecting soft sensors, control-advisory models) received lifecycle documentation (data/model cards, AD/UQ, monitoring) and defined human oversight. GPAI/foundation-model usage (literature assistants, coding copilots) gained supplier diligence and prompt governance.

What worked: early risk mapping (R1–R4), staged authority, and change-control integrated with the existing QMS.

7.5.2 Signal 2—Governing LLM Usage for SOPs and Literature

Organisations adopted templated prompts, retrieval-augmented search over approved corpora, and ban lists (no sensitive data, no execution instructions). All LLM outputs feeding SOPs or batch records required human review and attribution. Near-misses (ambiguous synthesis steps) reinforced the norm: LLM = assistant, not authority.

What worked: Prompt libraries, red-team tests for hallucination, and two-person sign-off for procedural changes.

7.5.3 Signal 3—Bias Audits in NAM-first Safety Portfolios

NAM-first toxicology groups ran coverage audits by chemotype and assay source. Models under-serving industrial chemistries triggered data acquisition or read-across supplementation. Dashboards tracked stratified performance and rejection rates, increasing regulator confidence.

What worked: Simple, routine by-stratum metrics and transparent AD rules.

7.5.4 Signal 4—Cyber Hygiene for Autonomous Lab Platforms

SDLs formalised network segmentation, code-signing for recipes, and edge fail-closed policies. Simulated latency incidents validated safe-halt behaviours and watchdog timers in fluidics/photo-chemical rigs.

What worked: Periodic chaos testing, explicit trip conditions, post-mortem learning.

7.5.5 Signal 5—Multi-objective Optimisation with Environmental KPIs

Plants extended BO/MPC objectives to include energy, PMI/E-factor, and NOx/CO_2, producing Pareto options for engineering and quality. Documentation tied decisions to company emissions pledges and permits—linking ethics to operations in measurable ways.

What worked: Governance that pays its way (quality and sustainability).

7.6 Common Pitfalls and How to Mitigate Them

7.6.1 "Great Metrics, Wrong Problem"

Pitfall: Optimising AUC on a dataset that does not reflect intended use (wrong chemotype/process/site mix).

Mitigation: Pre-register splits/strata; require by-stratum performance and coverage maps; keep an external test.

7.6.2 Opaque Decisions in High-stakes Contexts

Pitfall: Non-explainable model drives release or safety control.

Mitigation: Demand interpretability artefacts (attributions/counterfactuals), challenger models, and human sign-off.

7.6.3 Over-reliance on Single Surrogates

Pitfall: Docking score, one spectral proxy, or a single NAM result drives a regulatory claim.

Mitigation: Use ensembles/WoE, orthogonal evidence, and UQ; codify thresholds.

7.6.4 Authority Creep in Autonomy

Pitfall: "Advisory" models silently become de facto control.

Mitigation: Stage authority; log operator acceptance; periodically verify HMI behaviour against SOPs.

7.6.5 Unbounded LLMs in QMS

Pitfall: Draft SOPs/batch narratives without review; prompt leaks of confidential or personal data.

Mitigation: RAG over approved corpora; ban lists; human review with attribution; logging; privacy redaction.

7.6.6 Model Drift Unnoticed

Pitfall: Feedstock/instrument changes erode accuracy; alarms become noisy.

Mitigation: Drift monitors; re-calibration SOPs; retrain triggers; near-miss reviews.

7.6.7 Cybersecurity Gaps at OT/IT Boundary

Pitfall: Flat networks or unauthenticated protocol bridges.

Mitigation: Segmentation, MFA, signed recipes, edge inference, incident playbooks.

7.6.8 Documentation Debt

Pitfall: Models live only in notebooks; no seeds, no versions, no card.

Mitigation: Enforce model/data cards; version control; change-control gates.

7.7 Implementation Guidance (Step-by-step)

7.7.1 Risk Triage and Governance Plan (Two-hour Workshop)

Catalogue models—purpose, users, decisions, risk category (R1–R4).

Map frameworks—EU AI Act category (where relevant), ISO/NIST controls, sectoral duties (GxP/REACH/Part 11).

Assign owners—technical owner, process owner, executive sponsor.

Define artefacts—data cards, model cards, HAZOP-AI worksheet, monitoring dashboard.

Set gates—seven lifecycle gates (proposal → retire).

Decide authority—shadow → advisory → limited loop; safety envelope and overrides.

7.7.2   Minimum Documentation Pack (One Model)

Data card (sources, consent/legal basis, coverage maps, known biases, retention).

Model card (intent, data, splits, metrics by stratum, UQ/AD, interpretability, limits, retrain triggers).

Operating SOP (HMI, confidence display, escalation, overrides).

Monitoring SOP (KPIs, drift tests, reject rates, weekly/monthly review).

Change control (impact/risk assessment, validation plan, rollback).

Security note (interfaces, edge/cloud, encryption, credentials).

Ethics note (stakeholders, fairness checks, environmental KPIs).

7.7.3   Validating Models for Chemistry Decisions

Pre-specify metrics tuned to error cost (*e.g.*, false negatives for safety).

Use realistic splits (scaffold/time/site) plus an external test.

Report calibration (Brier/reliability); add conformal sets/intervals.[17]

Stratify performance (chemotype, site, instrument, community).

Demonstrate AD boundaries (latent/descriptor distance maps).

Show interpretability (alerts/AOP links; reaction or spectral features; counterfactuals).

Pilot in shadow/advisory modes; log adoption and overrides.

7.7.4   HAZOP-AI in Practice (One Page per Use-case)

Node: *E.g.*, "Soft sensor for residual solvent in dryer."

Deviations: Underestimates; delayed inference; stale model; OOD batch.

Causes: Feed composition change; sensor drift; pipeline failure.

Consequences: Release risk; rework; safety exceedance.

Safeguards: UQ/AD reject; Part 11-compliant interlock; MPC final control; hard constraints.

Actions: Safe-halt; lab confirmatory test; re-calibration; RCA.

Evidence: Card refs; validation; cybersecurity asset register.

7.7.5   Vendor/Foundation-model Diligence

Attestations: Training corpora, licences, safety filters, known limitations.

Benchmarks: Domain-specific tests (chemistry reading, SOP drafting, spectral reasoning).

Guardrails: RAG to approved corpora; prompt templates; no executable procedures without human review.

Logging: Prompts/outputs; attribution when content enters QMS.

7.7.6 Monitoring and Incident Response

Dashboards: Accuracy/calibration, reject ratios, AD breaches, fairness gaps, alarm loads, operator acceptance.

Near-misses: Capture and classify (data/label/pipeline/security); feed into weekly review.

Incidents: Define who triggers safe-halt; communications; regulatory notifications; blameless post-mortems; CAPA entries.

7.8 Key Takeaways

Proportionate governance: Scale controls to risk (R1–R4); build governance into the lifecycle.

Make uncertainty visible: Use conformal UQ and AD to bound decisions; route ambiguity to humans/assays.

Keep humans in charge:[19] Define safety envelopes, staged authority, and explicit overrides; log who/when/why.

Operationalise ethics: Stratified metrics, coverage maps, fairness checks, and environmental KPIs in optimisation.

Document relentlessly: Data/model cards, SOPs, change-control, audit trails—undocumented AI is not deployable.

Secure the edges: Segment OT/IT, sign recipes, prefer edge inference; maintain incident playbooks.

Be standards-literate: align with EU AI Act, ISO 23894/42001, NIST AI RMF, and sectoral rules (GxP/Part 11, REACH/CLP).

Iterate: monitor drift, near-misses, and adoption; refine models and processes continuously.

References

1. European Union, Regulation (EU) 2024/1689—Artificial Intelligence Act (Official Journal, 2024).
2. ISO/IEC, ISO/IEC 23894:2023—Information Technology—Artificial Intelligence—Risk Management.
3. ISO/IEC, ISO/IEC 42001:2024—Artificial Intelligence—Management System (AIMS).
4. NIST, AI Risk Management Framework 1.0, National Institute of Standards and Technology, 2023, DOI: 10.6028/NIST.AI.100-1.
5. OECD, Council Recommendation on AI (OECD/LEGAL/0449), 2019.

6. ISPE, GAMP 5 (2nd edn)—A Risk-Based Approach to Compliant GxP Computerized Systems, 2022.
7. U.S. FDA, 21 CFR Part 11—Electronic Records; Electronic Signatures, 1997/2010 updates.
8. ICH Q8(R2), Pharmaceutical Development, ICH, 2009.
9. ICH Q9(R1), Quality Risk Management, ICH, 2020.
10. ICH Q10, Pharmaceutical Quality System, ICH, 2008.
11. ICH Q14, Analytical Procedure Development, ICH, 2023.
12. IEC, IEC 62443—Security for Industrial Automation and Control Systems (series).
13. L. Floridi and J. Cowls, Five Principles for AI in Society, *Harv. Data Sci. Rev.*, 2019, **1**(1), DOI: 10.1162/99608f92.8cd550d1.
14. M. Mitchell, *et al.*, Model Cards for Model Reporting, *FAccT*, 2019, DOI: 10.1145/3287560.3287596.
15. T. Gebru, *et al.*, Datasheets for Datasets, *Commun. ACM*, 2021, **64**(12), 86–92.
16. M. T. Ribeiro, *et al.*, Why Should I Trust You?": Explaining Classifiers, *KDD*, 2016, DOI: 10.1145/2939672.2939778.
17. A. N. Angelopoulos and S. Bates, Conformal Prediction: A Gentle Introduction, *Found. Trends Mach. Learn.*, 2023, **16**(4), 494–591.
18. A. B. Arrieta, *et al.*, Explainable AI (XAI): Concepts and Challenges, *Inf. Fusion*, 2020, **58**, 82–115.
19. K. R. Varshney, Trustworthy Machine Learning in Engineering, *ACM Queue*, 2022, **20**(4), DOI: 10.1145/3546962.3560439.
20. ECHA, REACH—Guidance on Information Requirements and Chemical Safety Assessment (as of 2024).
21. European Commission, CLP Regulation (EC) No 1272/2008—Classification, Labelling and Packaging.
22. U.S. EPA, TSCA—Statutes & Guidance (as of 2024).

8 The Future of AI in Chemistry

8.1 Chapter Overview

What will chemistry look like when models understand matter, laboratories negotiate with algorithms, and computation is ambient, abundant, and audited? This chapter looks beyond current best practice to identify near-term inflections and durable trajectories. We focus on three forces reshaping the field: (i) foundation models and agentic systems that treat molecules, materials, proteins, spectra, and text as one continuous information space; (ii) cloud-native AI that lowers barriers to advanced computation and automation; and (iii) quantum–classical hybrids that aim to expand what is tractable in molecular modelling and reaction dynamics (Figures 8.1–8.3).

We ground the discussion in signals that already changed expectations in 2024–2025: AlphaFold 3's complex modelling across macromolecules and ligands; ESM3-class protein language models demonstrating sequence–structure–function design; the first peer-reviewed Phase 2a readout for an AI-designed small molecule; and scaled materials campaigns where models and robotics validated dozens of crystals within days–weeks. We then ask what these point towards: multimodal, constraint-aware foundation models for matter; self-driving R&D that is less a hero demo and more a platform; human–AI collaboration that is expert-centred and evidence-driven; and quantum-accelerated workflows where appropriate.

Rather than speculate broadly, we identify concrete methods and workflows chemists can adopt now to be "future-ready": data

RSC Foundations No. 7
AI Revolution in Chemistry
By Brian McKew
© Brian McKew 2026
Published by the Royal Society of Chemistry, www.rsc.org

Future Stack for AI-Native Chemistry

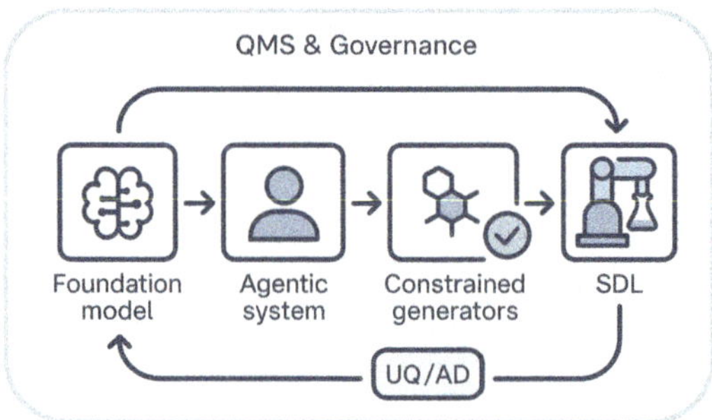

Figure 8.1 Future stack for AI-native chemistry.

Cloud-Native R&D Pipeline (Serverless

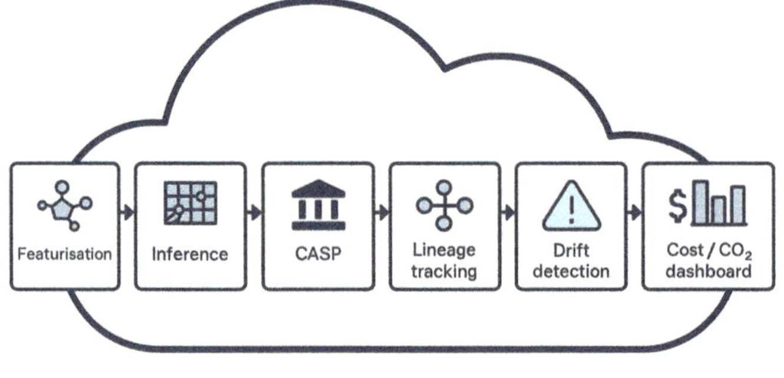

Figure 8.2 Cloud-native R&D pipeline (serverless).

Pragmatic Quantum–Classical Hybrid

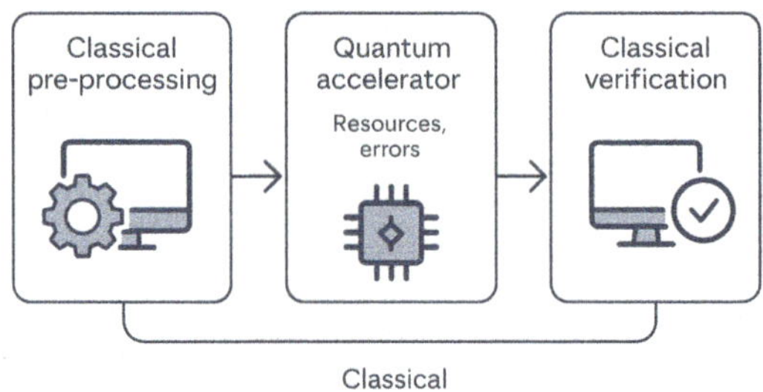

Figure 8.3 Pragmatic quantum–classical hybrid.

architectures for continual learning; evaluation harnesses for agentic tools; retrieval-augmented generation (RAG) over validated corpora; pattern-safe, guard-railed laboratory copilots; serverless chemistry pipelines; and hybrid quantum experiments with honest resource and error accounting. We close with pitfall-aware guidance—how to avoid over-claiming, vendor lock-in, compute sprawl, hallucinations, and governance debt—followed by checklists you can apply immediately.

8.2   Core Concepts and Definitions

8.2.1   Foundation Models for Matter[9]

Idea. A foundation model (FM) is trained at scale to learn reusable representations across tasks. In chemistry, emerging FMs ingest graphs (molecules/materials), 3D structures, sequences (proteins/ peptides/polymers), images/spectra, and text (protocols, literature). The aim is cross-modal competence: "given a target function and process constraints, propose candidates; explain the motifs; outline synthesis and characterisation; update beliefs from data".
   What changes.
   Unified embeddings: One latent space supports property prediction, inverse design, retrosynthesis, and spectral interpretation.
   Constraint conditioning: Synthesise-ability, green metrics, cost/ criticality, process windows, and safety become first-class inputs.
   Few-shot elasticity: FMs adapt to small datasets (new assay, substrate, or reactor) *via* fine-tuning or prompt conditioning.
   Explanation hooks: attention/attribution grounded in chemical priors (alerts, motifs, AOP KEs, reaction rules, lattice symmetry) make outputs discussable.
   Near-term reality. AF3-class complex models and ESM-class protein LLMs already raised baselines in structural reasoning; materials stacks (graph + diffusion) already propose stable crystals at scale. Expect convergence to multimodal, multi-scale FMs treating design, generation, docking/complex reasoning, spectra, and protocols as linked tasks with UQ/AD by design.

8.2.2   Agentic Systems and Autonomous R&D

Idea. An "agent" decomposes a goal into tool-using steps: search literature, propose candidates, call predictors, check constraints, write procedures, schedule experiments, interpret results, and update the plan. Orchestrated with robotics and PAT, agents form self-driving laboratories (SDLs).

What changes.

Throughput and cadence: DMTA cycles shrink to hours–days; process tuning uses multi-objective BO with feasibility guardrails (quality + energy + emissions).

Evidence-centred loops: Agents judged by calibrated UQ, AD, and value of information, not "creative" output alone.

Human–AI contracts: Chemists set briefs/constraints/stopping rules, review counterfactuals, approve escalations; agents remain bounded by safety envelopes.

8.2.3 Cloud-native AI in Chemistry

Idea. Cloud platforms provide on-demand accelerators (GPU/TPU), managed model hubs, serverless pipelines, data/version services, and API-first chemistry tools (retrosynthesis, procedure extraction, ELN/ LIMS connectors). This democratises access, supports reproducibility, and enables collaboration.

What changes.

Lower entry cost: spin up property-prediction, RAG assistants, or agentic pipelines in hours.

Standardised MLOps: model registries, feature stores, drift dashboards, model/data cards are table stakes.

Compute governance: track carbon/£ per experiment; set budget caps; prefer serverless to idle clusters; push edge inference near instruments.

8.2.4 Quantum–Classical Hybrids[13]

Idea. Quantum algorithms may accelerate parts of electronic structure/dynamics in the fault-tolerant era.[15] Near-term NISQ devices are limited; still, hybrid workflows—high-accuracy classical solvers plus variational quantum subroutines or quantum-inspired tensor methods—are worth targeted exploration.

What changes.

Scope: Early wins, if any, will be problem-specific (strongly correlated fragments, model Hamiltonians, excited-state subspaces) with rigorous resource and error tables.

Access: Quantum back-ends arrive as cloud endpoints; integrate as another accelerator in your pipeline.

Reality check. The future is hybrid: use quantum judiciously where it matters, surrounded by classical FMs, physics-informed ML, and HPC.

8.2.5 A Human-centred Future

Copilots for ideation, design, and analysis will be ubiquitous—but experts remain in charge. Productive labs nurture new roles: data stewards, model stewards, automation engineers, and prompt/RAG designers who keep assistants accurate, cited, and safe.

8.3 Methods and Workflows

8.3.1 Building a Future-ready Design Stack (Today)[8,12]

Design brief as code: Express design goals and constraints in a machine-readable brief (potency/selectivity windows; synthesise-ability/green constraints; process/quality/energy targets). Version in Git/ELN.

RAG over validated corpora: Index SOPs, assay methods, lab notebooks, reaction notes, and safety sheets. Use RAG so copilots cite approved content, not the open web.

Multimodal predictors: Combine graph (molecules/materials), sequence (proteins/polymers), and 3D modules;[10] always log UQ/AD and by-stratum metrics (chemotype/site/instrument).

Constraint-first generation: Fuse diffusion/RL with hard constraints (allowed reactions/building blocks, max steps, solvent/reagent exclusions, CRM/embodied-energy bounds).

Retrosynthesis + procedures: Send candidates through CASP; convert literature to action lists; emit robot/human-executable protocols; two-person review for safety-critical steps.

Agent orchestration: Run bounded loops: propose → predict → check constraints/UQ/AD → plan → (optionally) schedule → interpret → update. Surface counterfactuals, rationale ("why this, not that"), and stop on pre-set conditions.

SDL interface: Link agents to scheduling and PAT; use BO (single/multi-objective) with feasibility classifiers; hold control authority in MPC with hard constraints.

8.3.2 Cloud-native Chemistry Pipelines[6]

Serverless workflows. Orchestrate batch/event jobs (fingerprinting, featurisation, inference, scoring, CASP, scheduling), tracked in a model registry and feature store.

Data versioning and lineage: Tie results to dataset hash, split seed, model version, hyperparameters, code hash; retain raws for ALCOA+.

Cost and carbon governance: Tag jobs by project; present £ per kg of learning (cost per verified improvement); prefer spot/serverless; set quotas/alerts; track kg CO_2e.

Security posture: Segment networks; encrypt at rest/in transit; prefer edge inference; disable public endpoints; sign containers.

8.3.3 Evaluation Harness for Agentic Tools[7]

Task suites: Templated tasks (*e.g.*, "propose 10 routes under constraints X", "optimise photoredox yield with Y objectives").

Ground truth: Compare to curated baselines/lab results; score with task-specific metrics (success, violations, cost/time, E-factor/energy).

Hallucination safety: Require verifiable citations for procedural claims; penalise uncited assertions; never execute procedures verbatim without human review.

Robustness: Add adversarial cases (ambiguous prompts, missing metadata, conflicting signals) and measure graceful failure (reject/ clarify; never fabricate).

8.3.4 Hybrid Quantum Experiments (Pragmatic)[14]

Select cases: Strong correlation fragments, chemically relevant active spaces, toy Hamiltonians teaching future capability.

Resource estimates: Document qubits, depth, error budget, classical fallback.

Integrate: Treat quantum as an accelerator *via* API; cache intermediates; compare to CCSD(T)/DMRG where feasible.

Report honestly: Record wall-clock, queue time, cost, errors; flag approximations/limits.

8.4 Case Studies and Recent Signals (2024–2025)

Pattern recognition: Repeated signals that suggest durable shifts.

8.4.1 Signal A—Structure Reasoning Becomes "Complex-native"[1]

AlphaFold 3 extended from single chains to complexes (protein–ligand, RNA/DNA, ions, PTMs). Teams report faster cycles: mutants and ligand series prioritised with better contact hypotheses, then verified by orthogonal data. The pattern is not replacement but targeted experimentation.

8.4.2 Signal B—Programmatic Protein Design[2]

ESM3-class LLMs that link sequence–structure–function delivered working *de novo* proteins (*e.g.*, fluorescent variants) far from natural sequence space. Expect practical chemo-enzymatic routes *via* rapid variant ideation and few-shot screening guided by UQ.

8.4.3 Signal C—Clinical Signal for an AI-designed Small Molecule[3]

A TNIK inhibitor for idiopathic pulmonary fibrosis reported peer-reviewed Phase 2a results: acceptable safety/tolerability and dose-dependent lung-function changes over 12 weeks. The pipeline from AI ideation to human evidence is now copyable—and auditable.

8.4.4 Signal D—Scaled Materials Discovery with Autonomous Validation[4]

Graph/diffusion models proposed 10^5–10^6 crystals; A-Lab platforms synthesised/characterised dozens within days–weeks. The lesson is front-loaded pruning (composition/space-group/stability) plus closed-loop validation.

8.4.5 Signal E—SDLs as Capabilities[5]

2024 surveys documented SDL blueprints: modular robotics, PAT, BO/MPC, data/metadata spines, and safety envelopes. Standardisation—state machines, trip conditions, edge fail-closed—is replacing hero demos.

8.4.6 Signal F—Cloud R&D Becomes Routine

Hosted retrosynthesis/procedure extraction, model registries, and serverless pipelines shrank time-to-first-result and improved reproducibility (lineage). $Cost/CO_2$ dashboards now curb exuberance.

8.5 Common Pitfalls[11] and How to Mitigate Them

Over-promising from FMs $\rightarrow$ Constrain generation; require UQ/AD; maintain challenger baselines; demand orthogonal confirmation.

Vendor lock-in and compute sprawl $\rightarrow$ Prefer open formats and serverless; set budget/CO_2 caps; use multi-cloud abstractions.

Hallucination hazards $\rightarrow$ RAG over validated corpora; enforce citation checks; ban lists; two-person rule; "assistant, not authority".

Data debt and drift $\rightarrow$ Model/data cards; raw retention; drift monitors; re-calibration SOPs; pre-registered retrain triggers.

Ethics as paperwork $\rightarrow$ Publish coverage maps; track rejects by chemotype/site; link ethics to quality/energy/emissions KPIs.

Quantum cargo-culting $\rightarrow$ Require resource/error tables and classical baselines; publish limits and "why-quantum".

OT/IT security gaps $\rightarrow$ Segment networks; sign artefacts; prefer edge inference; enforce fail-closed; keep incident playbooks.

8.6 Implementation Guidance (Step-by-step)

8.6.1 Stand Up a "Future-ready" Project in 90 Days

8.6.1.1 *Week 1—Charter and Brief*

Define decision and design brief (objectives, constraints, safety envelope).

Pick 1–2 task suites; assign owners; open a QMS ticket with lifecycle gates.

8.6.1.2 *Weeks 2–4—Data and Baseline*

Build RAG over validated corpora; draft a data card; harmonise units/protocols; capture raws.

Train baselines (linear/trees/chemometrics); log UQ/AD; create coverage maps.

8.6.1.3 *Weeks 5–7—FM/Agent Integration*

Add multimodal predictors; enforce constraint-first generation; integrate CASP/procedure extraction.

Wire the agent harness (metrics, citation checks, counterfactuals).

If hardware is in scope, connect to scheduling/PAT; operate in shadow mode.

8.6.1.4 *Weeks 8–10—Close the Loop*

Run advisory cycles (design → predict → check → plan → test); compute value-per-experiment; compare to baselines.

Red-team ambiguity, missing metadata, conflicting sensors; record mitigations.

8.6.1.5 *Weeks 11–13—Operationalise*

Issue model/data cards, Operating/Monitoring SOPs; set drift/retrain triggers.

Review cost/CO_2 and security posture; stage path to limited closed loop (if applicable).

8.6.2 Hiring and Skills

Data steward, model steward, automation engineer (SDL/PAT/MPC), cloud engineer (serverless/MLOps), human-factors champion (HMI, explainability).

Upskill chemists in prompt/RAG design, error analysis, and UQ literacy.

8.6.3 Quantum Pilot Discipline

Choose one meaningful chemical case; prepare classical references; compute resource/error budgets; integrate *via* API.

Document outcomes and limits; decide go/hold on evidence, not enthusiasm.

8.7 Key Takeaways

Multimodal, constraint-aware foundation models will unify prediction, generation, and planning; insist on UQ/AD and explanations.

Agentic systems will turn objectives into experiments—keep them bounded, cited, and audited; humans decide.

Cloud-native pipelines reduce time-to-impact; govern cost/CO_2, security, and reproducibility *via* lineage.

Quantum will be hybrid and selective; require resource estimates and classical baselines.

Governance enables speed: data/model cards, drift monitoring, change control, safety envelopes.

Invest in people and skills: stewards, automation/cloud/MLOps, human factors; upskill chemists in UQ and RAG.

Start small, guard-rail heavily, and let evidence expand scope.

References

1. J. Abramson, *et al.*, Accurate structure prediction of biomolecular interactions with AlphaFold 3, *Nature*, 2024, **627**, 795–803.
2. T. Hayes, *et al.*, ESM3: sequence–structure–function frontier model, *Science*, 2025 (press/journal coverage).
3. Z. Xu, *et al.*, Generative AI–discovered TNIK inhibitor for IPF: Phase 2a trial, *Nat. Med.*, 2025, **31**, 2602–2610.
4. A. Merchant, *et al.*, Scaling deep learning for materials discovery, *Nature*, 2023, **624**, 80–87.
5. G. Tom, *et al.*, Self-Driving Laboratories for Chemistry and Materials Science, *Chem. Rev.*, 2024, **124**, 5905–5977.
6. P. Schwaller, *et al.*, Molecular Transformer: uncertainty-calibrated reaction prediction, *ACS Cent. Sci.*, 2019, **5**, 1572–1583.
7. P. Schwaller, *et al.*, Retrosynthetic disconnections and reaction conditions with transformers, *Mach. Learn.: Sci. Technol.*, 2021, **2**, 015016.
8. A. Dunn, Q. Wang and A. M. Ganose, *et al.*, Matbench: benchmarking materials property prediction, *npj Comput. Mater.*, 2020, **6**, 138.
9. C. Chen, W. Ye, Y. Zuo, C. Zheng and S. P. Ong, MEGNet framework, *Chem. Mater.*, 2019, **31**, 3564–3572.
10. T. Xie and J. C. Grossman, Crystal Graph Convolutional Neural Networks, *Phys. Rev. Lett.*, 2018, **120**, 145301.
11. K. M. Jablonka, *et al.*, Bias, fairness, and robustness in materials ML, *Chem. Mater.*, 2023, **35**, 4662–4677.
12. P. I. Frazier, A tutorial on Bayesian optimisation, *arXiv*, 2018, 1807.02811.
13. Y. Cao, *et al.*, Quantum chemistry in the age of quantum computing, *Chem. Rev.*, 2019, **119**, 10856–10915.
14. J. Tilly, *et al.*, Variational Quantum Eigensolver: a review, *PRX Quantum*, 2022, **3**, 030201.
15. J. Lee, *et al.*, Efficient quantum chemistry on fault-tolerant hardware, *PRX Quantum*, 2021, **2**, 030305.

9 Practical Implementation Guide

9.1 Chapter Overview

This chapter is a hands-on manual for chemists and pharmaceutical scientists who are ready to implement AI in their research or operations. If earlier chapters showed what is possible, here you will find how to start, how to evaluate, how to validate, and how to keep things running without compromising safety, quality, or scientific integrity. The guidance is designed to be used as you work: checklists, decision trees, and templates that plug into your existing lab notebooks, ELNs (electronic lab notebooks), LIMS (laboratory information management systems), QMS (quality management systems), and project rhythms.

You will learn how to frame a crisp decision statement that an AI system can actually help with; how to audit and prepare data; which representations (descriptors, graphs, sequences, 3D, spectra, images) match which problems; how to choose and benchmark baseline *vs.* advanced models; how to design validation that predicts real-world performance; when to run Bayesian optimisation (BO) or adopt self-driving lab (SDL) elements; and how to quantify and communicate uncertainty. You will also find troubleshooting guides for common failure modes—data leakage, confounding, "great retrospective, poor prospective," hallucinating assistants, uncalibrated soft sensors—and go/no-go criteria that help you escalate with confidence.

RSC Foundations No. 7
AI Revolution in Chemistry
By Brian McKew
© Brian McKew 2026
Published by the Royal Society of Chemistry, www.rsc.org

Throughout, we keep this practical: the target reader is a chemist who can install a Python environment and work with an ELN, but who may be new to model governance, MLOps, or edge deployment. The chapter closes with step-by-step playbooks you can apply to property prediction, reaction optimisation, materials screening, safety/toxicology QSAR, and process soft sensing—plus a compact set of templates (data card, model card, operating SOP, monitoring SOP) that will satisfy most reviewers and auditors.

9.2 Implementation Quick-start: Five Decisions to Unlock in Week One

9.2.1 Define the Decision and Action

Example: "Rank 500 analogues to pick 24 to make next, maximising potency and permeability while keeping TPSA 50–90 $\mathring{A}^2$ and $c \operatorname{Log} P$ 1–3."

9.2.2 Choose the Validation You Intend to Pass

Pick scaffold/time/site splits that mirror deployment; reserve an external test set from day one.

9.2.3 Baseline First

Commit to a simple, auditable baseline (regularised linear/logistic, random forest/gradient boosting, or chemometric PLS for spectra) before touching advanced learners.

9.2.4 Plan Uncertainty and Applicability Domain (AD)

Decide how you will compute and display UQ (conformal prediction/ ensembles) and how you will reject out-of-domain cases.

9.2.5 Pick the Escalation Path

Define the thresholds for moving from baseline to advanced models, and from manual experiments to BO/SDL elements—plus the safety envelope if any autonomy is involved.

9.3　Data Audit and Preparation (Checklist)

9.3.1　Structure and Identity

- Canonicalise structures; fix stereochemistry/tautomers; strip salts/solvents; unify charge state (*e.g.*, pH 7.4 for DMPK tasks).
- De-duplicate across vendors/ELN; ensure unique IDs; map merges to provenance.

9.3.2　Labels and Protocols

- Harmonise units/SOPs; record assay conditions (cell line, media, temperature/time, internal standards).
- Quantify noise (replicate variance); decide conflict-resolution policy (majority vote/quality-weighted).

9.3.3　Splits and Leakage Controls

- Pre-register splits (scaffold/time/site); forbid plate/operator/time identifiers in features unless causally justified.
- Hold out an external test set; lock it until final evaluation.

9.4　Representations: Mapping Problems to Encodings

9.4.1　Molecules and Materials

- Descriptors/fingerprints (baselines); GNNs for non-linear structure–property; add 3D only when shape/pose matter.[6,7]

9.4.2　Reactions and Procedures

- Templates/template-free CASP; action-list encodings for procedure extraction and automation.[4,5]

9.4.3　Spectra and Images

- Chemometrics for baselines; CNN/transformers where justified; always retain link to bands/peaks for interpretation.

9.4.4 Proteins and Complexes

- Sequence models for embeddings; AF-class structures for complex reasoning; docking/physics as orthogonal checks.

9.5 Model Selection and Benchmarking

9.5.1 Baselines

- Linear/logistic, random forests/gradient boosting, PLS for spectra—fast, interpretable, strong with small–medium data.

9.5.2 Advanced

- GNNs/transformers/diffusion when data and problem complexity warrant; record seeds/libraries/hyperparameters.

9.5.3 Metrics and Calibration

- Report ROC-AUC/PR-AUC (classification) or MAE/RMSE (regression); add calibration (Brier score, reliability plots).[8]
- Use conformal prediction for per-sample intervals/sets; verify coverage on validation folds and external test.

9.6 Validation that Predicts Deployment

- Use splits that match deployment (scaffold/time/site); stratify metrics by chemotype/site/instrument; track reject rates.
- Keep an untouched external test; do not tune on it.
- For BO/SDL, add prospective validation: pre-declare success criteria (*e.g.*, $\geq 80\%$ of top-k predicted actives confirmed; $\geq 10\%$ yield gain at $\leq 10\%$ energy increase).

9.7 When to Use BO and SDL Elements

9.7.1 BO

- Use when experiments are expensive and the design space is bounded (factors with safety/feasibility constraints).
- Include replication (10–20%) to measure noise and drift; use uncertainty-aware acquisition (EI/UCB) and diversity controls.[1,2]

9.7.2 SDL (Bounded Autonomy)

- Start with one instrument/one objective; connect model → propose → execute → measure → learn under safety envelope (hard limits, interlocks, safe-halt).[3]
- Keep authority staged: shadow → advisory → limited closed loop.

9.8 Uncertainty and AD in Operations

- Display UQ (intervals/sets) with every prediction; provide human-review routes for ambiguous cases.
- Define AD *via* descriptor/latent-space distance; log OOD rejections and coverage maps; plan targeted data acquisition to expand the domain.

9.9 Troubleshooting: Fastest Routes to Green

9.9.1 Leakage and Confounding

- Symptom: Great retrospective, poor prospective.
- Fix: Enforce scaffold/time/site splits; scrub plate/operator/time artefacts; run error forensics.

9.9.2 Hallucinating Assistants

- Symptom: Uncited or unsafe procedural claims.
- Fix: RAG over approved corpora; enforce citation checks; never execute unreviewed procedures.

9.9.3 Uncalibrated Soft Sensors

- Symptom: Drift after maintenance or feed change.
- Fix: Add calibration standards, drift monitors; schedule re-calibration; implement alarms on bias.

9.9.4 Objective Mismatch

- Symptom: Optimising a proxy that fails at QC.
- Fix: Align objective with released CQAs; add penalties (energy, solvent, off-spec cost).

9.10 Playbooks (Step-by-step)

9.10.1 Property Prediction (Molecules/Materials)

1. Data card (sources, SOPs, coverage, biases).
2. Baselines (descriptors/fingerprints $\rightarrow$ linear/trees); log UQ/AD.
3. Advanced (GNN/transformer) only if clear gain on deployment-like splits.
4. Calibration (reliability/Brier) + conformal intervals; stratify metrics.
5. External test once; freeze model; write model card.
6. Monitoring: drift, reject rate, by-stratum performance.

9.10.2 Reaction Optimisation (BO)

1. Define factors/bounds and safety limits (*e.g.*, $T \leq 120$ °C; solvent exclusions).
2. Seed with space-filling/DoE; set batch size (8–16).
3. Acquisition with diversity; replicate 10–20%; stop on convergence or budget.
4. Robustness: $\pm\Delta T$, $\pm10\%$ equivalents, different operator/lot/day.
5. Document gains and conditions; update SOPs.

9.10.3 Materials Screening

1. Two-stage pipeline: composition-only baselines $\rightarrow$ crystal-graph models for survivors.
2. Add stability filters (formation/decomposition energy); sustainability constraints (CRMs, embodied carbon).
3. Prospective validation in small batches; track UQ/AD coverage.

9.10.4 Safety/Toxicology QSAR

1. OECD-aligned: defined endpoint, algorithm, AD, performance, mechanistic anchors.[9]
2. Baselines first; advanced models only with documented gain.
3. Conformal/ensemble UQ; reject OOD; provide alerts/AOP links.[10]
4. Model card + WoE narrative.

9.10.5 Process Soft Sensing

1. Align PAT timelines with process tags; compute engineered windows/features.[11]

2. Baseline chemometrics; escalate to ensembles/CNNs only if justified.
3. Calibrate; define AD; add alarms for bias/drift; plan re-calibration.
4. Advisory use first; integrate with MPC under change control.[12]

9.11 Templates (Insert in Your QMS)

9.11.1 Data Card (One Page)

- Sources, SOPs, coverage maps, biases, units, rights/consents, retention.

9.11.2 Model Card (Two–Four Pages)

- Intent, data references, splits, metrics (overall/by stratum), calibration plots, AD definition, UQ method, limits, retrain triggers.

9.11.3 Operating SOP

- HMI, confidence display, escalation routes, overrides.

9.11.4 Monitoring SOP

- KPIs (accuracy/calibration, reject rate, drift), review cadence, retrain triggers.

9.11.5 Change-control Form

- Impact assessment, validation scope, rollback plan, approvals.

9.12 Key Takeaways

- Start every AI initiative by writing a one-sentence decision and action, then align data, models and validation to it.
- Build simple, auditable baselines first; only add BO, SDL or foundation models when they deliver clear deployment gains.
- Design validation, UQ and AD to mimic deployment conditions, with explicit reject rules and prospective tests where it matters.
- Treat playbooks (property prediction, optimisation, screening, QSAR, copilots, SDL, quantum) as reusable templates, not rigid recipes.

- Capture decisions, assumptions and results in lightweight data/model cards and SOPs so you can defend outcomes under scrutiny.
- Use governance artefacts (risk register, change control, monitoring plan) to unlock faster iteration, not to slow projects down.
- In week one, focus on the five decisions in this chapter rather than tools or vendors; technology choices can follow.

References

1. P. I. Frazier, A Tutorial on Bayesian Optimisation, *arXiv*, 2018, 1807.02811.
2. B. Shahriari, K. Swersky, Z. Wang, R. P. Adams and N. de Freitas, Taking the Human Out of the Loop: A Review of Bayesian Optimisation, *Proc. IEEE*, 2016, **104**, 148–175.
3. F. Häse, L. M. Roch and A. Aspuru-Guzik, Next-Generation Experimentation with Self-Driving Laboratories, *Trends Chem.*, 2019, **1**, 282–291.
4. C. W. Coley, *et al.*, Computer-Assisted Synthesis Planning: Status and Challenges, *Acc. Chem. Res.*, 2018, **51**, 1281–1289.
5. P. Schwaller, *et al.*, Molecular Transformer: Uncertainty-Calibrated Reaction Prediction, *ACS Cent. Sci.*, 2019, **5**, 1572–1583.
6. T. Xie and J. C. Grossman, Crystal Graph Convolutional Neural Networks, *Phys. Rev. Lett.*, 2018, **120**, 145301.
7. C. Chen, W. Ye, Y. Zuo, C. Zheng and S. P. Ong, MEGNet, *Chem. Mater.*, 2019, **31**, 3564–3572.
8. A. Dunn, Q. Wang, A. M. Ganose, D. Dopp and J. A. Matbench, *npj Comput. Mater.*, 2020, **6**, 138.
9. OECD, Principles for the Validation, for Regulatory Purposes, of (Q)SAR Models. OECD, 2004.
10. U. Norinder and S. Boyer, Conformal Prediction for Reliable Toxicity Classification, *J. Chem. Inf. Model.*, 2017, **57**, 459–468.
11. M. Easton, *et al.*, Real-Time Release Testing: Principles and Case Studies, *J. Pharm. Innov.*, 2021, **16**, 673–690.
12. C. Rogers, *et al.*, Digital Twins in Pharmaceutical Manufacturing, *Int. J. Pharm.*, 2022, **624**, 121–137.

10 Conclusion: Embracing the AI-driven Future of Chemistry

10.1 Chapter Overview

Artificial intelligence (AI) is no longer a side project in chemistry; it is a new instrument—one that records, learns and proposes at the speed of computation while remaining bounded by chemical reality. Across this primer we have explored what AI can do (predict, generate, plan, optimise), where it changes the tempo (hit finding,[3] materials search,[5] process control, toxicology), how to deploy it responsibly (validation, uncertainty, governance), and what is arriving next[1] (multimodal foundation models, agentic systems, cloud-native pipelines, pragmatic quantum–classical hybrids).

This final chapter synthesises the main learning points into a practical roadmap for continuing education and implementation. It shows how chemists can actively shape AI's evolution—by curating better data, insisting on interpretability and calibrated uncertainty, embedding sustainability and safety as first-class constraints, and building teams and processes that are fast and auditable. You will find a compact, repeatable pattern: decision first → data ready → baselines before brilliance → validation that predicts deployment → prospective confirmation → governance by design. We close with curated resources to help you keep current as the field moves (Figures 10.1–10.3).

RSC Foundations No. 7
AI Revolution in Chemistry
By Brian McKew
© Brian McKew 2026
Published by the Royal Society of Chemistry, www.rsc.org

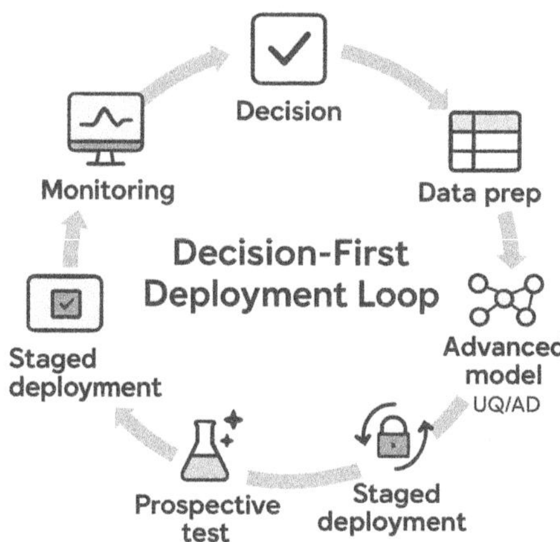

Figure 10.1   Decision-first deployment loop.

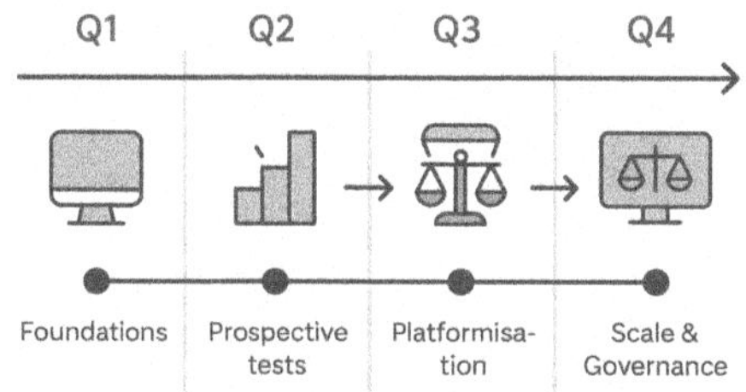

Figure 10.2   12-month roadmap for an AI-ready lab or plant.

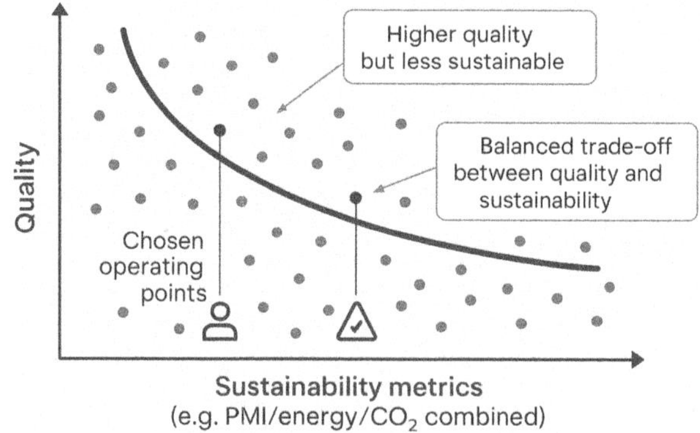

Figure 10.3   Sustainability-aware optimisation.

10.2 Core Concepts and Definitions

10.2.1 AI as an Instrument: Decision-centred Chemistry

Treat every AI application as a support for a specific decision that a chemist, team, or plant will take. Example decisions: "Rank catalysts to maximise isolated yield under green-solvent constraints," "Select 12 analogues to test for hERG risk at $\leq 1\%$ false-negative rate," "Recommend dryer set-points to reduce energy 5% while holding residual solvent within spec." The quality of the decision statement determines whether your models can be judged fairly and improved quickly.

10.2.2 Representations[5] Connect Data to Questions

Descriptors/fingerprints are strong baselines for small-molecule QSAR and screening.

Graphs/3D capture local environments and shape when SAR is non-linear or pose-dependent.

Sequences/structures represent proteins/polymers; crystal graphs represent periodic solids.

Spectra/images (IR/Raman/NMR, SEM/TEM) require standardised pre-processing and raw retention (ALCOA+).

Text should be mediated *via* retrieval-augmented generation (RAG) over approved corpora (SOPs, ELNs, institutional libraries), not the open web, for safety-critical content.

10.2.3 Validation that Predicts Deployment

Validation must mimic how the model will be used:

Use scaffold splits for medicinal chemistry SAR; time/site splits for processes; cluster splits for diverse QSARs; keep an external test untouched.

Report stratified metrics by chemotype, site, instrument, and substrate class; show calibration (reliability/Brier).

Quantify uncertainty (conformal sets/intervals; ensembles) and define applicability domain (AD) with explicit reject rules.[9]

10.2.4 Prospective Confirmation and Value of Experiment[16]

The performance that matters is prospective: a small, well-chosen experiment (16–48 compounds; 8–16 BO points) that proves the model's utility compared with baselines and heuristics. Track value per experiment: yield gain per synthesis, hit-rate lift, energy saved per trial, false-alarm reduction, or time saved per decision.

10.2.5 Governance that Unlocks Speed

Model cards, data cards,[11] change control,[10] and monitoring SOPs are enablers. They reduce dispute costs, accelerate audits, and allow safe iteration. Pair shadow $\rightarrow$ advisory $\rightarrow$ limited closed loop authority staging with safety envelopes (hard constraints, interlocks, safe-halt states).

10.2.6 Sustainability, Safety and Ethics as Constraints[14]

Encode synthesise-ability, green chemistry metrics (*E*-factor/PMI, embodied energy/carbon), critical raw materials avoidance, and safety exclusions inside generation, BO and control—not as after-the-fact filters. For toxicology, integrate hazard (QSAR, read-across, NAMs) with exposure/kinetics *via* PBPK/IVIVE to deliver risk curves.

10.2.7 Human–AI Collaboration

Chemists remain the decision-makers. Agents and copilots propose, summarise and schedule; chemists set briefs and constraints, review counterfactuals, sign off changes, and lead error forensics. The most productive labs cultivate data stewards, model stewards, automation engineers, and RAG/prompt designers.

10.3 Methods and Workflows

10.3.1 Method 1—The Decision-first Loop[4] (Repeatable Pattern)

Frame the decision and success thresholds (pre-registered).

Prepare data: Cards; units/protocol harmonisation; structural normalisation; spectral pre-processing (baseline, SNV/MSC, derivatives); time alignment for processes.

Establish baselines: Linear/regularised, tree ensembles, chemometrics (PCA/PLS); publish a clear business floor.

Escalate selectively: GNNs/transformers/3D only if warranted by endpoint and data; keep baselines as challengers.

Validate like deployment: Scaffold/time/site/cluster splits; stratified metrics; calibration; UQ/AD and reject rules.

Confirm prospectively: Mini-batch experiment; compute value per experiment; run perturbation/robustness checks.

Stage deployment: Shadow $\rightarrow$ advisory (confidence/AD in HMI) $\rightarrow$ limited closed loop under constraints; monitor KPIs and alarms.

Operate and learn: Drift monitors, near-miss logs, retrain triggers; periodic ethics/fairness reviews; publish coverage maps.

10.3.2 Method 2—Education that Compounds (for Individuals and Teams)

Core skills for all chemists: Decision framing; data literacy (units, protocols, batch effects); reading calibration plots; understanding UQ/AD; writing a one-page model card; RAG usage and citation checks.

Specialist roles:

Data steward—metadata/lineage, dataset hygiene, coverage maps.

Model steward—model cards, calibration/UQ, AD maps, change control.

Automation engineer—SDL, PAT, BO/MPC, state machines, safety envelopes.

Cloud/MLOps engineer—serverless pipelines, registries, feature stores, cost/CO_2 dashboards.

Human-factors lead—HMI design, interpretability, operator adoption.

Learning cadence: monthly error-forensics seminars; quarterly red-team drills (ambiguous prompts, missing metadata, conflicting sensors); annual refreshers on standards (ISO/IEC 42001; NIST AI RMF; GAMP 5).[12,13]

10.3.3 Method 3—Evaluation Harness for Agentic Tools[6] and Copilots

Task suites: Curated prompts with objective scoring (success rate, violation count, cost/time, *E*-factor/energy).

Citation discipline: Penalise uncited assertions; never execute generated procedures verbatim; two-person review for SOP changes.[7]

Robustness probes: Deliberately ambiguous inputs; missing fields; contradictory signals. Expect graceful failure (ask for clarification, decline) rather than fabrication.

Hallucination firewall: RAG over approved corpora; safety ban-lists (reagents, conditions); sandboxed outputs.

10.3.4   Method 4—Sustainability[15,17] by Design

Objective augmentation: Extend objectives with energy, PMI/E-factor, NOx/CO_2, criticality. Use EHVI for multi-objective BO.[8]

Data: Supply-chain metadata, embodied energy/carbon databases, solvent guides.

Reporting: Present Pareto sets with annotated sustainability; record decisions with their trade-off rationale.

10.3.5   Method 5—Pragmatic Quantum–Classical Pilots

Case selection: Chemistry where strong correlation or excited states resist classical shortcuts.

Resource and error budgets: Qubits, depth, readout error, wall-clock, queue time, £/run; classical baselines (CCSD(T), DMRG).

Integration: Quantum as an accelerator behind an API; cache intermediates; compare to classical references.

Publication: Emphasise limits; avoid cargo-culting.

10.4   Key Takeaways

Own the decision—clarity of action, constraints, and thresholds drives impact.

Validate like you deploy—realistic splits, stratified metrics, calibrated UQ/AD, and a prospective mini-batch.

Govern to go faster—cards, change control, and monitoring allow rapid, safe iteration.

Constrain for reality—synthesise-ability, safety and sustainability belong inside generators, BO, and control.

People make it work—stewards, automation and cloud engineers, and human-factors champions multiply returns.

Agents assist; humans decide—bounded, cited, audited assistance with counterfactuals and orthogonal confirmation.

Think platform, not demo—bounded loops that deliver value become rails for autonomy.

Stay curious—and honest—track cost/CO_2; publish limits; expand scope on evidence.

References

1. J. Abramson, *et al.*, Accurate structure prediction of biomolecular interactions with AlphaFold 3, *Nature*, 2024, **627**, 795–803.
2. T. Hayes, *et al.*, ESM3: a frontier model of sequence–structure–function; design of *de novo* proteins, *Science*, 2025, **387**, 850–858.
3. Z. Xu, *et al.*, Generative AI–discovered TNIK inhibitor for IPF: randomised Phase 2a trial, *Nat. Med.*, 2025, **31**, 2602–2610.
4. G. Tom, *et al.*, Self-Driving Laboratories for Chemistry and Materials Science, *Chem. Rev.*, 2024, **124**, 5905–5977.
5. A. Merchant, *et al.*, Scaling deep learning for materials discovery, *Nature*, 2023, **624**, 80–87.
6. P. Schwaller, *et al.*, Molecular Transformer: uncertainty-calibrated reaction prediction, *ACS Cent. Sci.*, 2019, **5**, 1572–1583.
7. P. Schwaller, *et al.*, Predicting retrosynthetic disconnections and reaction conditions with transformers, *Mach. Learn.: Sci. Technol.*, 2021, **2**, 015016.
8. P. I. Frazier, A tutorial on Bayesian optimisation, *arXiv*, 2018, 1807.02811.
9. A. N. Angelopoulos and S. Bates, Conformal Prediction: A Gentle Introduction, *Found. Trends Mach. Learn.*, 2023, **16**(4), 494–591.
10. M. Mitchell; *et al.* Model Cards for Model Reporting. *FAccT* 2019, DOI: 10.1145/3287560.3287596
11. T. Gebru, *et al.*, Datasheets for Datasets, *Commun. ACM*, 2021, **64**(12), 86–92.
12. ISO/IEC 42001:2024, Artificial Intelligence—Management System (AIMS), https://www.iso.org/standard/81230.html.
13. NIST, AI Risk Management Framework 1.0. 2023, DOI: 10.6028/NIST.AI.100-1.
14. OECD, Guidance on (Q)SAR Validation and AOP/PBK resources (various).
15. B. I. Escher, *et al.*, From chemical mixtures to effect-based water quality monitoring, *Environ. Sci. Technol.*, 2018, **52**, 9767–9775.
16. P. Raccuglia, *et al.*, Machine-learning-assisted materials discovery from failed experiments, *Nature*, 2016, **533**, 73–76.
17. Y. Chen, *et al.*, Inverse design of solid electrolytes with targeted ionic conductivity using ML, *Energy Environ. Sci.*, 2021, **14**, 6157–6168.

Further Reading

1. J. Abramson, *et al*, Accurate structure prediction of biomolecular interactions with AlphaFold 3, *Nature*, 2024, **627**, 795–803.
2. P. Schwaller, *et al*, Molecular Transformer: Uncertainty-calibrated reaction prediction, *ACS Cent. Sci.*, 2019, **5**, 1572–1583.
3. M. H. S. Segler, M. Preuß and M. P. Waller, Planning chemical syntheses with deep neural networks and symbolic AI, *Nature*, 2018, **555**, 604–610.
4. A. Merchant, *et al*, Scaling deep learning for materials discovery, *Nature*, 2023, **624**, 80–87.
5. G. Tom, *et al*, Self-Driving Laboratories for Chemistry and Materials Science, *Chem. Rev.*, 2024, **124**, 5905–5977.
6. T. Xie and J. C. Grossman, Crystal Graph Convolutional Neural Networks, *Phys. Rev. Lett.*, 2018, **120**, 145301.
7. C. Chen, W. Ye, Y. Zuo, C. Zheng and S. P. Ong, MEGNet: A universal graph deep-learning framework, *Chem. Mater.*, 2019, **31**, 3564–3572.
8. A. Dunn, Q. Wang and A. M. Ganose, *et al*, Matbench: benchmarking materials property prediction, *npj Computational Materials*, 2020, **6**, 138.
9. P. Raccuglia, *et al*, Machine-learning-assisted materials discovery from failed experiments, *Nature*, 2016, **533**, 73–76.

RSC Foundations No. 7
AI Revolution in Chemistry
By Brian McKew
© Brian McKew 2026
Published by the Royal Society of Chemistry, www.rsc.org